Wild Animals
OF NORTH AMERICA

Wild Animals
OF NORTH AMERICA

NATIONAL
GEOGRAPHIC
SOCIETY

Wild Animals
of North America

Preceding Pages: Panorama of North American wildlife profiles an Alaskan brown bear haloed by a rainbow in Katmai; a potbellied marmot surveying its lush Olympic National Park domain; a humpback whale streaming waterfalls in Glacier Bay, Alaska; a regal elk greeting the morning sun in Nebraska's Fort Niobrara National Wildlife Refuge.

Published by
The National Geographic Society

Gilbert M. Grosvenor
*President and
Chairman of the Board*

Michela A. English
Senior Vice President

William R. Gray
*Vice President and
Director, Book Division*

Prepared by National Geographic
Book Service
Charles O. Hyman, *Director*

Editorial Consultant
Ronald M. Nowak
Author of Walker's Mammals
of the World, 5th edition

Chapters by Dr. Nowak *and*

Howard W. Campbell
*National Fish and Wildlife
Laboratory, Gainesville, Florida*

Joseph A. Chapman
*College of Natural Resources,
Utah State University*

Alfred L. Gardner
*Patuxent Environmental Science
Center, National Biological Service*

Valerius Geist
*Environmental Design,
University of Calgary*

Hugh H. Genoways
*Museum Studies Program,
University of Nebraska*

Maurice G. Hornocker
*Hornocker Wildlife Research
Institute, Moscow, Idaho*

Charles Jonkel
*Ursid Research Center,
Missoula, Montana*

Karl W. Kenyon
*U. S. Fish and Wildlife Service
(retired)*

L. David Mech
*Department of Fisheries and
Wildlife, University of Minnesota*

John L. Paradiso
*Office of Endangered Species,
U. S. Fish and Wildlife Service
(retired)*

Victor B. Scheffer
*Former Chairman,
U. S. Marine Mammal Commission*

Stephen R. Seater
Wildlife Biologist

Merlin D. Tuttle
*Bat Conservation International,
Austin, Texas*

Ralph M. Wetzel
*Department of Biology,
University of Connecticut*

Staff for this Book

Thomas B. Allen
Editor

Shirley L. Scott
Assistant Editor

David M. Seager
Art Director

Anne Dirkes Kobor
Illustrations Editor

Linda B. Meyerriecks
Asst. Illustrations Editor

Molly Kohler
Illustrations Research

Jules B. Billard
Mary H. Dickinson
Seymour L. Fishbein
Edward Lanouette
Margaret Sedeen
Verla Lee Smith
Editor-Writers

Wayne Barrett
Mary Swain Hoover
Robert M. McClung
Elizabeth L. Newhouse
David F. Robinson
Elizabeth C. Wagner
Edward O. Welles, Jr.
Writers

Susan C. Eckert
Suzanne P. Kane
Penelope A. Loeffler
Allison Rogers
Karen Hoffman Vollmer
Anne Elizabeth Withers
Research

Richard S. Wain
Andrea Crosman
Leslie A. Adams
Production

John T. Dunn
Quality Control Director

David V. Evans
Engraving and Printing

Karen F. Edwards
Traffic Manager

Lise M. Swinson
Editorial Assistant

Jeffrey A. Brown
Index

Contributions by
Greta Arnold
Ross S. Bennett
Melanie A. Corner
Effie M. Cottman
Henri A. Delanghe
Charlotte J. Golin
M. Washburn Swain

First edition 400,000 copies
Second printing 25,000 copies
Revised edition (1987) 200,000 copies
Revised edition (1995) 65,000 copies
Library of Congress æ data page 406

Contents

Animals for All Seasons

A white-tailed doe nuzzles her suckling. The very substance of mammal—mother's milk—bonds the pair. The fawn will stay close to its mother for the first year of its life.

A WILD ANIMAL LIVES beyond human control and support. It finds its own food and shelter, and nature alone determines the time for mating and the bearing of young. Insects are animals, as are worms, fish, frogs, snakes, and birds. But when many of us say "animals" we really are thinking of "mammals," the group of usually furred and four-footed creatures to which we ourselves belong.

More than 470 species of wild mammals find ways to live in the varied habitats of the Nearctic Zoogeographical Realm—the lands and their adjacent waters that stretch from Canada's Arctic islands to Mexico's central highlands. For many of these mammals, survival is a daily struggle, often intensified by human presence.

The word "mammal" comes from the Latin *mamma,* which refers to the breast or nipple through which the female passes milk to her young. All mammals have at least one pair of nipples, except for the primitive, egg-laying monotremes—the platypus and spiny anteaters of Australia and New Guinea. But even these animals have mammary glands, and their young lap milk as it seeps through pores in the abdominal skin.

Nourishment of offspring by the mother's milk unites all mammals and distinguishes them as a group from other kinds of animals. Nursing also helps to explain some other aspects of the mammal's way of life. The direct dependence of the young on the body of the mother, even after the young are born, tends to lead to an extended period of affinity between individuals, and hence to the development of social relations, communication, and the transmission of knowledge from one generation to another.

The sound of the word *mammal* is essentially the same sound made by a human infant in association with its natural source of nourishment. And this sound has been incorporated around the world into the word for the female parent: *ma* in Chinese, *maht* in Russian, *madre* in Spanish, *mutter* in German, and *mother* in English.

If maternal care is one feature that wins some sympathy from us for our fellow mammals, another is the presence of the soft, attractive body covering that we call hair or fur. True hair, a nonliving derivative of the outermost layer of skin, is found only in mammals. Most species have both an underfur, consisting of a dense layer of short, fine hairs; and an array of longer, coarser guard hairs that extend above the underfur and usually contribute most to the color and pattern we see. Hairs are shed either throughout the year or all at once in annual or seasonal molts.

Hair's primary function is insulation. A covering of hair traps air and retards the loss of body heat. Combined with oily secretions from glands, hair helps to waterproof the body. Hair also helps an animal conceal itself. The coloration of most mammals, unlike that of many birds, is drab and lets the animal blend with the vegetation, soil, or light conditions. Some species, including certain North American hares, lemmings, and weasels, have brownish summer fur, which is replaced in the autumn molt by white fur for camouflage against snow. Hair also acts as a shield against skin injury, as a tactile device in the form of whiskers, as a defensive weapon when modified into quills or spines, and as a means of communication when formed into patterns that can be displayed to others.

Mammals that lack a covering of fur usually have gained other advantages. Sparsely haired armadillos, for example, have developed hardened skin for protection. No mammal goes an entire lifetime without some hair,

Grunting his ardor, a bighorn ram approaches a ewe. In his courtship—frequently part of mammalian reproduction—he rears and kicks as he pursues the ewe up wooded cliffs. To assert mating rights, he bangs heads with rivals.

Mock fighting stimulates ocelots prior to mating. Playfully they nip each other about the head. As the rite continues, the female's behavior changes from spitting and clawing to purring and rubbing.

though whales and porpoises may have hair only in the embryonic stage or, as adults, they may have just a few bristles around the mouth. For insulation, these sea creatures have evolved a layer of blubber which in some species may be up to 24 inches (60 cm) thick.

An insulating coat is one of several features that enable mammals to follow a more vigorous, adaptive life than that of their reptilian ancestors. Another is the heart, which in mammals (and birds) is fully separated into left and right sections. Thus, when freshly aerated blood from the lungs enters the left side of the heart to be pumped throughout the body, the blood cannot mix with the deoxygenated blood returning to the right side of the heart for pumping back to the lungs.

Mammals are also helped by the presence, between the mouth and nasal passages, of a hard secondary palate that allows simultaneous breathing and chewing.

A more efficient oxygen supply to the tissues means a higher metabolic rate and more available energy. With the heat that results, and with insulation that retains the heat, body temperature is maintained at a relatively constant level, regardless of conditions outside the body. Many mammals may thus survive in frigid climates, function on cold nights, and run for long distances.

In general, mammals rid themselves of excess heat through sweating or panting, while reptiles seek shade. But, even with temperature-control mechanisms, some

mammals regularly find themselves unable to produce enough body heat from available food. One solution to the problem is hibernation. In this state of extreme dormancy or torpor, body temperature may plummet to near freezing and heartbeat and respiration drop to a small fraction of normal levels; the animal can be awakened only with difficulty.

Among North America's mammals, true hibernation is limited to certain bats and various rodents, including ground squirrels,

For energy, an animal must eat. And, through the specialized development of their teeth, mammals have attained an advantage over other animals in securing and processing food. While a reptile's teeth, for example, generally look all alike, most mammals are heterodont, meaning their teeth are differentiated so that they can cut, slice, crush, and grind food. Thus the food is delivered into the body in a form that can be easily digested and used to maintain high energy levels.

One mother, one young: A harbor seal, fattening on rich milk, may double its weight in four weeks. Soon after she weans the pup the mother is ready to breed again. She normally bears a single pup each year.

woodchucks, and jumping mice. Some other species, such as bears, skunks, and the raccoon, sometimes spend much of the winter sleeping in their dens and living off reserves of fat. But this slumber involves no drastic reduction in metabolism.

Habits vary. Chipmunks, for example, may be dormant throughout the winter, only for a few of the coldest days, or not at all. Certain bats become torpid every day, while they roost, to conserve energy for flying at night.

Most mammals have four types of teeth. Proceeding from the front to the back of the mouth, they are the incisors, canines, premolars, and molars (page 24). Usually there are uppers and lowers of each kind.

Incisors are usually small teeth used for nipping bits of food. But in rodents and rabbits, the incisors have become large, powerful gnawing instruments which, unlike most mammalian teeth, continue to grow throughout life. (The elephant's tusks are specialized incisors, helpful in obtaining food by bending branches and prying up bark.)

Canines, not present in rodents or rabbits and absent or relatively small in many larger plant-eating mammals, have become long and powerful in most mammals that capture other animals for food. The largest canines, the tusks of the walrus, seem to function primarily as weapons.

Premolars take several forms. Meat-eating mammals may have premolars that resemble the edge of a saw and are used for cutting. In plant eaters they sometimes are difficult to distinguish from molars, which usually are broad, with flat crowns bearing a number of cusps or ridges. The pattern formed on the crown varies between groups and helps in tracing evolutionary lines. Molars, which mainly crush and grind food, achieve their greatest development in plant-eating species.

The evolution of advanced dentition in mammals was associated with development of

One mother, abundant young: Nursing deer mice cling to their mother's six mammae, holding on even if she suddenly moves from the nest. She whelps three or four litters in a year. The young mature sexually within six weeks.

a stronger, more efficient arrangement of the jaws. In fish, amphibians, reptiles, and birds, each side of the lower jaw is composed of several different, pieced-together bones; this jaw does not connect directly with the skull. In the course of evolution, two of the bones of each side of the reptilian lower jaw became parts of the mammalian middle ear. Each side of the mammal's lower jaw came to consist of a single bone, articulating (hinging) with the skull. This evolutionary architecture made the jaws a powerful engine for the food-processing teeth.

No matter how well the jaws and teeth function, they are useless without the means of getting to the food—and then, if need be, getting safely away. In the evolution from reptile to mammal, limbs came to extend downward from the body, rather than outward from the sides, thus facilitating more rapid movement. But some mammals—people included—still are relatively slow, and these species usually walk by placing the entire foot, from heel to toe, on the ground.

Many mammals to which speed is life— such as the wild dogs and cats—have feet with the heel raised well above the ground. Deer, horses, and many of the other big plant eaters, which must run for long distances, move about on the tips of their toes and have developed large, protective nails—hoofs.

Some mammals have feet highly modified for special means of locomotion. Rabbits and hares have large hind feet which, together with smaller front feet, allow a rapid leaping gait. The kangaroos of Australia—and their nonrelated namesakes, the kangaroo rats and mice of North America—have, relative to the body, even larger hind feet that provide for bounding without the aid of forelegs. Moles and gophers, which spend much of their lives tunneling through the soil, have big and powerful front feet and claws.

To escape danger or to search for food, many mammals can climb, aided by powerful limbs and flexible, well-developed digits and claws. Squirrels are our most arboreal mammals. The so-called flying squirrels can glide—they do not actually fly—by opening a loose fold of skin that extends from wrist to ankle. The only truly flying mammals are bats, which use their power to chase insects or reach fruits and flowers.

Most kinds of mammals can swim, and many routinely cross small bodies of water. Some, such as beavers, muskrats, and river otters, are, with their webbed feet and flattened tails, expressly adapted to a semiaquatic life. The tail has become short on the sea otter, which spends most of its life in the water. But its hind feet have expanded into flippers. The pinnipeds—seals, sea lions, and walruses— have a rudimentary tail, and all four feet have become finlike. The tail of the manatee has been modified into a large paddle, while the hind limbs have disappeared. Whales and porpoises also lack hind limbs, and have become so adapted for movement through water that they sometimes are mistaken for fish. Their bodies are streamlined, and horizontal, unboned flukes have developed on their tails to provide propulsion.

Complex activity requires control by a highly evolved nervous system. The brains of mammals, especially the cerebral hemispheres, are relatively much larger than those of other animals. These hemispheres form a memory center, allowing new sensations to be evaluated on the basis of past experience. From this phenomenon came the growth of what we call intelligence.

Their mental and physical features have enabled mammals to occupy nearly all parts of the world and adapt to a multitude of challenges. Could any environment on earth be more inhospitable than the ice-covered Arctic Ocean? Yet the polar bear thrives there, drifting on the ice hundreds of miles from solid land, swimming through the frigid waters. Polar bears probably would do fine in parts of Antarctica, if they could get there. But habitat conditions along the way would not be conducive to their kind of life.

The habitat of any animal is the space that it can utilize to support itself. Some mammals get along well under a great variety of

Born to climb, a young goat tests its heritage under mother's eye. Young mammals often learn the ways of their species by imitation of elders and in play.

OVERLEAF: *Born to jump, a kangaroo rat springs on elongated hind legs. The multiple exposure shows its tufted tail helping to brake for a landing. Jumping is one of the ways mammals get around. As they evolved, they developed varied locomotion: from walking and running to swimming and flying.*

Antlers held high, caribou wade Arctic waters. Herds tens of thousands strong make the most spectacular migrations of North American land mammals. They travel hundreds of miles between their seasonal ranges.

conditions. The little deer mouse ranges over much of North America, largely unrestricted by variations in altitude, temperature, and plant life. Its relative, the Florida mouse, however, is found solely in Florida—and only in areas of sandy soil with sparse vegetation. Certain meat eaters manage to live in many kinds of habitats because their only main requirement seems to be the presence of enough other mammals as prey.

Habitat condition affects the population density of mammals, and smaller species generally have higher densities. Among many species of rodents, there are about two to twenty individuals per acre, but there is seldom more than one raccoon for every five acres. The lowest densities are among large predatory mammals, such as the gray wolf, which even under the best conditions averages only about one per ten square miles.

The niche of a species does not refer to its living space but to the role that the species plays in nature, relative to the resources used, and the time, location, and rate of such use. Niches seldom are the same, though they may overlap. Professor Tom Kunz of Boston University once found that at least six species of bats hunted insects each night in the same general areas of central Iowa. But the various species tended to avoid excessive competition by hunting at different times of the night or in separate localities.

Surprisingly few North American mammals are strongly diurnal—meaning active mainly during daylight. Although many species are occasionally seen at midday, they usually move about and seek food at night (nocturnal species) or during twilight hours (crepuscular species).

There are exceptions, including some shrews, bison, bighorn sheep, and most of the squirrel family. Arboreal squirrels require good light conditions to avoid accidents as they leap about the trees. Flying squirrels, however, have excellent night vision and the ability to glide; they are exclusively nocturnal. Ground-dwelling members of the squirrel family are mainly diurnal, but they usually

remain near an extensive burrow system where they can find refuge from predators. This need for safety is probably one reason why so many other mammals move mostly in the dark. Some of them may also seek to avoid the heat of the day.

Mammals usually must adjust their activity to the season. Some mammals, including many predatory species, must work harder and travel farther in winter to secure enough food. Deer, moose, caribou, and other large hoofed

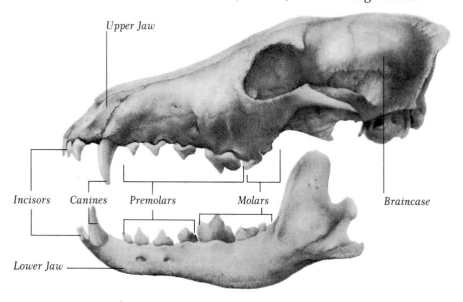

Upper Jaw

Incisors Canines Premolars Molars

Braincase

Lower Jaw

A coyote's skull has the typical signs of a mammal: a massive, one-piece lower jaw and a set of highly diversified teeth.

mammals often shift their ranges in winter and concentrate in places where they can find suitable vegetation. When such mass movements become regular annual journeys that cover substantial distances, they are called migrations.

The area normally used by an animal, when it is not in migration, is its home range. This area may be less than half an acre (0.2 ha) for a mouse or as large as 5,000 square miles (13,000 sq km) for a wolf pack. In winter, the size of a home range generally contracts for plant eaters and expands for meat eaters.

Most mammals use particular places for shelter, rest, safety, or giving birth to young. This characteristic seems to be least apparent among large hoofed species, which may do

nothing more than occasionally conceal themselves in thickets. But among the rodents we find the elaborate burrows of prairie dogs, the suspended nest of golden mice, and the complex den of beavers.

Within the home range of many kinds of mammals is a certain area that an individual or a social group prefers to keep exclusively for itself. Such areas, staunchly defended against at least some other members of the same species, are called territories. The area defended may correspond closely with the entire home range, but often the area is much smaller. Territoriality probably serves to give adequate space to individuals with respect to available resources. Or, as in the case of certain seals and sea lions, territoriality mainly involves the defense of a small spot by a male during the breeding season. Within this area he forms a harem, attempting to control as many females as possible.

Mountain lions mark large territories with scent and visual signs, seeming satisfied never to see others of their kind, except for mating and maternal care. Virginia opossums apparently have no territories. But they also try to avoid one another—and fight savagely when they cannot.

In many other species, social activity has developed beyond a purely antagonistic stage. While the woodchuck is a solitary creature, its western cousin, the Olympic marmot, is a tolerant, sociable rodent that forms large colonies based on a mated pair and several generations of offspring.

Establishment of bonds between a male and female, and sometimes their young, is one way in which mammalian social life has grown. Social expansion also occurs through bonds between females. A Richardson's ground squirrel mother may allow her female offspring to settle nearby and ultimately inherit her burrow and territory. In some species the bond is so strong that several generations of female offspring eventually accumulate. This is a basis of group formation in species as diverse as mountain sheep, sperm whales, and coatis.

Young males, usually less amicable than females, sometimes form their own groups. A pronghorn population consists of herds of bachelors, as well as female herds and territorial males. The bachelors establish a dominance hierarchy to maintain order among themselves. But there comes a time when harmonious relationships give way to the urge that brings forth future generations. Then the young may challenge the old for territory and the right to mate.

Shrews, rabbits, mice, and voles often can mate and bear young in the same season in which they are born. But male marmots, wolves, sperm whales, and pronghorns usually must wait until long after they are physiologically ready. Such is the cost of life in an organized group where older and stronger individuals already are established.

Because of the dangers of the wild, probably fewer than half of all mammals even survive to mate and only a tiny fraction live to full potential. Maximum known longevity is less than a year for some shrews to more than 100 years for the fin whale.

One individual's lifetime forms only a small phase of evolution, the process through which change and diversification occur in living things. Biologists now recognize that evolution has led to the development of several great kingdoms of life. One, the animal kingdom, contains numerous groups, including the vertebrates, animals with backbones. Seven classes of living vertebrates are known: Agnatha (lampreys and hagfishes), Chondrichthyes (sharks, skates, and rays), Osteichthyes (other fish), Amphibia (frogs, salamanders), Reptilia (turtles, lizards, snakes), Aves (birds), and Mammalia.

Zoologists, in an effort to show relationships, break these classes into progressively smaller groups: orders, families, genera, and species. Sometimes they throw in infraclasses, superorders, and subfamilies. Of these groups, only the species is fully a natural unit. For only at the species level do the animals themselves recognize the relationship and behave (Continued on page 30)

The animals of this book live in the Nearctic Zoogeographical Realm, one of the world's six such major realms. Each region and adjacent waters supports its own distinctive and indigenous forms of animal life, as well as those it shares with other realms.

The Nearctic (new north) extends from the Aleutian Islands to Greenland and reaches south into the highlands of central Mexico. There the Nearctic blends into the Neotropical Realm of Central and South America. More than 470 mammalian species inhabit the Nearctic. Some others have been introduced deliberately or by accident to the realm.

A list of native mammals appears on pages 394 to 397.

Chiroptera

Lagomorpha

Cetacea

Artiodactyla

Mammals: in a class by themselves

Like a map of the mammals, this chart shows the journey of a great class through time—and through scientific terminology. Some names may be unfamiliar, but they point the way to how North America got its mammals.

Biologists divide the animal kingdom into several groups, or phyla, which in turn are separated into classes. Among animals with backbones—vertebrates—there are three classes for fish, and one each for amphibians, reptiles, birds, and our class: mammals.

The first mammals probably evolved about 200 million years ago from reptiles that had developed some characteristics usually associated with mammals.

At some remote stage, mammals separated into at least two great subclasses, Prototheria and Theria. (Features of their skulls distinguish members of the two groups.) Each subclass, in turn, is divided into three infraclasses.

Of the Prototheria, only one infraclass survives—and only in the form of the duck-billed platypus and spiny anteaters of the order Monotremata. They are restricted to Australia and New Guinea. Unlike all other living mammals, they reproduce by laying eggs. Two other infraclasses, once widely distributed, are now extinct.

Of the subclass Theria, two major infraclasses still exist. One, the Metatheria, includes a single order of

fascinating animals: the marsupials. They give birth to incompletely developed young that usually are raised in a maternal pouch.

All other existing mammals are in the infraclass Eutheria. The success of this group may result from possession of a well-developed placenta, a membrane system that allows embryonic young to be nourished from the mother's body more efficiently than in the marsupials—and thus be born in a fairly advanced state.

With the extinction of the dinosaurs and most other orders of reptiles some 65 million years ago, mammals began to extensively diversify. They developed varied ways to get around, find food, and

otherwise survive. They thus expanded to fill vacant niches and adapt to lifestyles never followed by other kinds of animals.

Mammals of the infraclass Eutheria have evolved into 17 orders, as shown on the diagram. Of these, 10 are native to our Nearctic Realm. (Three others, Perissodactyla, Proboscidea, and Primates, once lived in North America.) In the chart's boxes, each Nearctic order is symbolized by a physical characteristic.

OVERLEAF: From prairie dog to black bear—a sampler of the mammals of North America.

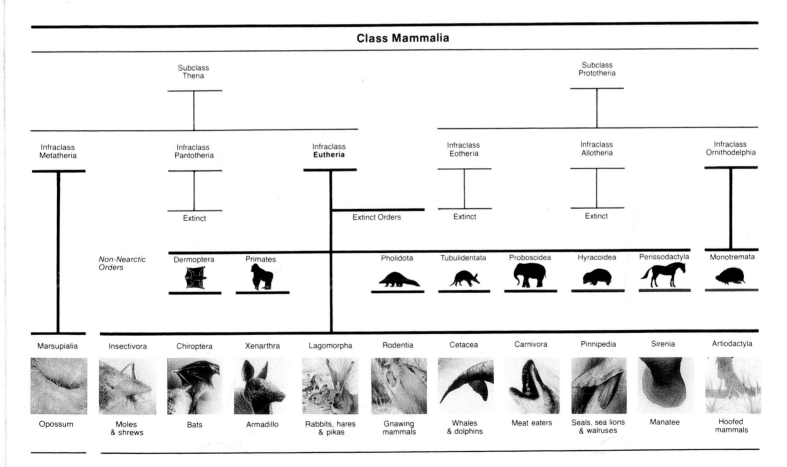

Insectivora

Rodentia

Sirenia

Marsupialia

Pinnipedia

Carnivora

Xenarthra

On a land no longer their own, the big and the small still seek a way of life—the prairie dog to burrow, the buffalo to roam.

(Continued from page 25) accordingly. A species is usually thought of as a group of individuals capable of interbreeding only among themselves and producing offspring that have the same capability.

Within the cells of the members of a species are particles called genes, which carry the instructions for assembly of future generations. Genes are spread and recombined through sexual union. The physical changes wrought by genetic modifications become incorporated within some or all populations of a species. Acceptance of modifications depends in part on how well the changes are suited for the environment.

For a new species to be formed, there must be both variation and reproductive isolation. An established rodent species, for example, might be widely distributed but still be united by gene flow. One population of the species might then be isolated by movement of a glacier, development of a desert, or formation of some other barrier. Since genetic flow would no longer be possible, this population might evolve in a different way, especially if environmental conditions on its side of the barrier underwent a great change.

In time the isolated population might become physically incompatible with its parent species, even if the barrier disappeared. No interbreeding between the two populations would occur—and a new species would have come into existence.

Biologists give every species a two-part Latin name, the first representing the genus and the second the species. A genus may contain only one species or several closely related species. Thus *Sciurus* is the generic name for a number of arboreal squirrels found over much of the world. *Sciurus carolinensis* applies only to the common gray squirrel of eastern North America. Sometimes a third Latin word is added to the name to designate a subspecies or geographic race.

North America's native species once included giant ground sloths, horses, tapirs, camels, mastodons, and mammoths. Their predators were lions, saber-toothed cats, massive dire wolves, and giant bears. All became extinct around 12,000 to 8,000 years ago. Two entire orders—Perissodactyla (which included the horse and tapir) and Proboscidea (mammoth and mastodon)—were then suddenly eliminated in North America. But members of eleven orders did survive, and we live amid them today.

The sequence of orders in this book begins with those orders whose members are thought to most closely resemble ancestral mammals. Orders that differ the most from early mammals are placed last. Among the first are the shrews and moles, for they apparently are little changed from the earliest placental mammals. The artiodactyls—the deer and the bison are among them—come last because they are so different from the early mammals. They have, for instance, greatly modified limbs that allow these hoofed mammals to run from predators at high speed for long distances.

At the very beginning, however, is the opossum, because its order—Marsupialia—can be traced farther back in time than any of the living placental orders, and also because marsupials have biological features that set them apart from all other mammals.

So turn the page and meet the opossum—first in a presentation of the mammals that enrich North America and, by their survival, give us wonders to study or simply to observe and enjoy.

RONALD M. NOWAK

The Opossum

ORDER Marsupialia

Many a marsupial shared the land with the dinosaurs. Today only one pouched mammal ranges north of Mexico—the Virginia opossum, a young species that first appeared during the Ice Age.

IN THE YEAR 1500, Vicente Yáñez Pinzón made two major discoveries: Brazil and, on its shores, a marsupial. The animal was a female opossum carrying young. Yáñez Pinzón, who had shared in another major discovery eight years earlier as captain of the *Niña* under Columbus, took this "incredible mother" back to Spain. There the monarchs Ferdinand and Isabella were among the first Europeans to marvel at its pouch, that "sensational contrivance of Nature."

The name by which we best know marsupials comes from an Indian word, *apasum*—white animal. It was introduced to Western civilization in 1608, in Captain John Smith's description of the animal from Virginia: "An Opassum hath an head like a Swine, and a taile like a Rat, and is of the bignes of a Cat. Under her belly she hath a bagge, wherein shee lodgeth . . . and sucketh her young."

Scientists of the day believed marsupials to be unique to the New World. A century later the East Indian filander and cuscus were recognized as marsupials. And then the 1768-71 expedition led by Captain James Cook brought news of the astounding variety of the Australian marsupials. The name opossum was applied to most of them.

Recognizing this diversity, mammalogists have recently begun to divide marsupials into several full orders. The order that would include the opossum under this scheme is Didelphimorphia.

The significance of the unique method of reproduction common to all marsupials went unappreciated until the 19th century. The two wombs, the common external opening for reproductive and digestive systems, and the embryonic state of the tiny newborn identify the Marsupialia; the order at present includes about 280 species. Marsupial gestation is brief—as short as 12 days in some species. At birth each grain-size young must make a precarious search for its mother's nipples. The newborn that succeeds firmly attaches to a teat, where it completes the developmental process.

The abdominal pouch, or marsupium, the "sensational contrivance" that first drew special attention to the group, is not found in all species. Among those that do have a pouch, some develop it only during the breeding season. Nor is the marsupium unique to marsupials; it occurs in the egg-laying spiny anteaters and duck-billed platypus as well.

Marsupials dominate the Australian and Tasmanian fauna and spill over into New Guinea and the Celebes and Moluccas. In addition to the familiar kangaroos, wallabies, and koalas, the diverse fauna there includes marsupial counterparts of our moles, shrews, flying squirrels, and wolves. Some mouselike species weigh as little as a 25-cent piece. The largest, the red kangaroo, weighs nearly 200 pounds (91 kg) and stands more than 6 feet (1.8 m) tall. These living giants are dwarfed by their extinct cousins, whose remains indicate a rhinoceros-size bulk.

Elsewhere, marsupials occur only in the New World. They abound in the tropical forests of South America, though not with the variety of species found in Australia. Nevertheless, they include flesh-eating "shrew" opossums, tree-dwelling woolly opossums, and the ground dwellers whose light spots above the eyes explain their name—the four-eyed opossums. The yapok, which haunts streams from Mexico to Brazil, resembles the otter in habits. It has webbed feet and a special muscle that closes the pouch watertight.

Most New World opossums are rat-size. The largest are in the group that includes the

Playing possum: a life-and-death game. Curled body, open eyes, grisly grin, and lax tail typify the pose. When danger threatens and there's no escape, the Virginia opossum may sink into this nervous paralysis, a catatonic state that can last just a few minutes or several hours.

Regional culture celebrates the opossum in song and folklore. It is trapped for fur, hunted for sport and food. "Possum and taters" is a southern tradition. Many are kept as pets. Rules for the selection of "Top Possum" at an Alabama fair required all entries to wear an official leash. They had to be free of parasites; any that bit a judge twice faced automatic disqualification.

For all its fame, the opossum often puzzles suburbanites in a backyard encounter. Best clues to its identity: bare ears and tail.

cat-size Virginia opossum. Old males may weigh more than 12 pounds (5.4 kg), though average weight is closer to half that. Length averages 2.9 feet (88 cm).

Marsupial and placental mammals share a common ancestry and represent different adaptive trends that arose by mid-Cretaceous times, about 100 million years ago. We are not sure where the marsupials originated. Some paleontologists favor North America because of its rich fossil record of marsupials from late Cretaceous times. Fewer kinds are known from South American deposits of similar age—some 70 or 80 million years old. But if the South American Cretaceous were as well known as that of North America, paleontologists might favor the Southern Hemisphere as the probable site of marsupial origin. In Australia, which has the most species, the fossil record is a blank until the late Oligocene, some 30 million years ago.

The great distance separating the Australian region from the Americas is not the zoogeographic anomaly it once seemed. Today's geological evidence indicates that Australia, Antarctica, and South America, along with Africa and the Indian subcontinent, once formed a single continental landmass called Gondwanaland. About 150 million years ago Gondwanaland began to break up into the units we know today. Each gradually drifted toward its present location.

When marsupials appeared, Australia, Antarctica, and South America were still associated and likely shared some of the same animals. Australia broke away and was isolated by an oceanic moat during the evolution of most of its marsupials.

But North and South American marsupials had to contend with highly competitive placental mammals. By Miocene times, which began 26 million years ago, North American marsupials had lost the struggle for survival. By then, some marsupials had invaded Europe, Africa, and possibly Asia, but none of them survived.

The only marsupials inhabiting North America belong to the family Didelphidae. Of these, only the Virginia opossum occurs in the Nearctic Realm, the area covered by this book. Our opossum probably evolved in Mexico late in the Ice Age; today it ranges from Costa Rica to Canada. Its ancestor, *D. marsupialis*—the common opossum of tropical America—is the same species captured by Yáñez Pinzón in Brazil. This opossum of the tropics is no match for the Virginia opossum in the drier, colder habitats of northern Mexico and the United States.

Archaeological evidence shows that the Virginia opossum was rare or absent north of present-day Virginia and Ohio when Europeans began colonizing this continent. Today the species is in all New England states and southern Canada. But it is probably at the extreme of tolerable winter climates. Opossum ears may be lost and the tail frozen back to a stub during severe winters in the northern parts of the range. Canadian populations are periodically exterminated by harsh winters, then replaced from the south in milder years. Populations in the far west were all begun from animals brought to the region by people.

Biologists have long considered marsupials to be primitive hangers-on from earlier times, inferior to placentals in intelligence, adaptability, and reproductive efficiency. This view is being challenged as we learn more about marsupials. The greatest difference between the two groups is in reproduction. Some researchers now interpret the marsupial method as providing the female greater control over the rearing of her young because her only major commitment of energy— production of milk and carrying, care, and protection of the brood—occurs after they are born and not before.

The marsupial brain is relatively small, but not substantially different in structure from that in other mammals. Experiments in learning have shown that marsupials are not the "intellectual inferiors" of placentals. In the face of competition from modern placentals, the Virginia opossum increases its range and numbers. It even thrives amid 20th-century civilization.   ALFRED L. GARDNER

— Range Boundary c. 1900
— Range Boundary c. 1600

As human population increased in the North, so did the opossum, aided by new food sources: garbage and road kills. The species first got to the West around 1870 when fans transplanted some to California.

Virginia Opossum *(Didelphis virginiana)*

As John Smith noted, the opossum is about the size of a house cat, although heavier bodied and shorter legged. The long guard hair is usually gray, but black-phase animals are common in the South. The opposable big toe and the female pouch appear in no other animal covered in this book.

The pouch has 12 teats arranged in a horseshoe arc around a 13th in the center. A litter that may number 20 or more is born after a gestation of $12\frac{1}{2}$ to 13 days. Most die; 8 is the usual number found in the pouch. Blind and naked, the bee-size newborn clamber with their strong forelimbs along the mother's freshly washed hair from

vulva to fur-lined pouch. Once attached to nipples, the babies maintain their hold for some 60 days. Thereafter, they can leave and regain the nipple at will, suckling until about 100 days after birth.

The Virginia opossum usually lives in woodland; look for its star-shaped tracks near streams and swamps. Nocturnal by habit, it

Like an astronaut tethered to his capsule, a baby opossum clings to an elongated nipple in mother's pouch. Stretched as the young grow, these lifelines allow freedom of movement for a crowded litter. After a couple of months the young can let go and may ride mother's back until fully weaned.

If the need arises, opossums can travel a tree limb upside down. Hind feet, with an opposed big toe, grip like a human hand. The tail can grip too; in this "behind hand" the opossum can haul woodland debris to cozy up a sleeping den.

often appears slow and clumsy by day. It does not hibernate, though activity slows in winter. Breeding generally occurs in late winter and again in early summer. Leaf-lined holes, hollows, and crevices serve as dens.

Food consists mostly of grasses, nuts, and fruits—but anything that happens to walk, hop, or crawl by may make a meal. The opossum is reputedly a chicken thief, and also is fond of snakes, including deadly pit vipers. It prowls trash cans and city dumps and—at great risk—scavenges highway carrion.

Many scientists think the term *Didelphis* refers to the double reproductive tract (*di*—two, *delph*—womb). When the name was first used, however, the internal anatomy was unknown. *Didelphis* referred to the internal womb and the pouch, considered an external uterus.

Among folk beliefs still with us is the idea that the male copulates with the female's nose. This arises from the fact that the male has a forked penis and the only visible paired openings in the female are her nostrils. Also, during birth she licks clean the pouch and the fur between it and the vulva. People figured she was blowing embryos from her nose into the pouch.

Shrews & Moles

ORDER Insectivora

OCCASIONALLY DESCRIBED as the "wastebasket" of the mammal class, this order forms a convenient catchall for certain mammals that zoologists aren't quite sure where else to put. Insectivores, divided into seven living families, are found worldwide except in Australia, Greenland, the polar regions, and the southern half of South America. They range from little-known West Indian solenodons and Old World tenrecs and hedgehogs to the more familiar shrews and moles. Only the last two live in North America.

Scientists generally agree that the "insect eaters" resemble the earliest placental mammals that lived on earth over 70 million years ago. They gave rise to the more advanced mammals, including monkeys and apes to which humans are allied. Of all insectivores, shrews have probably stayed closest in form and habits to the ancestral group.

All American insectivores have long, slender, sensitive snouts and small rudimentary eyes. Their feet have five clawed digits, and their dense fur generally lacks the long guard hairs characteristic of many mammals. Some have a strong, musky odor that may deter predators. None of them hibernates. Insatiably hungry, they are on the go almost constantly, searching for insects, worms, and other small animal life. Sometimes they eat plant matter such as roots and seeds.

Most insectivores are small. In fact, one of the world's smallest mammals, in terms of weight, is the American pygmy shrew, which weighs less than one-tenth of an ounce. It is fortunate for us that shrews and moles are no larger. Insectivores are among the most ferocious and pugnacious of animals. If shrews were the size of tigers or lions, they would be formidable predators.

Because of their rapid rate of metabolism, shrews must eat frequently or risk dying of starvation. Some consume more than their own weight in food each day. Their voracious hunger allows for no niceties. Animal behaviorist Konrad Lorenz watched in dismay as several baby water shrews in his laboratory began to devour a large frog alive: "The poor frog croaked heartrendingly, as the jaws of the shrews munched audibly in chorus." He hastily ended the experiment.

Teamwork is not the norm after infancy. Shrews are fierce loners and appear to be territorial. Residents defend their home ground against invading shrews with a repertoire of threatening postures, loud squeals, and, if necessary, furious attack.

The shrew's ferocity counts for little against its natural predators—hawks, weasels, and snakes, among others. Its underground relative, the mole, fares better by staying mostly in its burrow.

Most of us are familiar with moles from their habit of digging tunnels that disrupt gardens, manicured lawns, and golf courses. Although their burrowing does sometimes harm the roots of delicate flowers and vegetables, they perform a useful service by aerating and loosening soil.

They tunnel with amazing strength and speed. Eastern moles dig shallow runways at a rate of 10 to 20 feet (3 to 6 m) an hour—up to 160 feet (48 m) in one night. To match a mole, a man would have to excavate in the same time a tunnel nearly half a mile long and large enough to wriggle through.

Moles, like shrews, eat Japanese beetles and their larvae, cutworms, wireworms, and other insects harmful to crops and garden plants. Given the huge number of these ravenous insect eaters, it is clear that they play an important part in maintaining nature's balance, and gardeners should be glad to have them as allies.
JOHN L. PARADISO

Family Soricidae

When the family cat proudly deposits a dead shrew at the door, we usually mistake the victim for a mouse. For shrews—small, grayish or brownish in color, with long, pointed snouts—do superficially resemble mice. Although seldom seen or recognized by the average person, they rank among our most numerous animals. But habitat destruction potentially endangers them. Thirty-three species live in the Nearctic, from rain forests to deserts and from sea level to the highest mountains.

While they may burrow a little, shrews are not really adapted for digging. Instead they dart swiftly over the ground, or just under the leaf litter in which they live, rooting for insects and other small prey. Water shrews can run over water. Hair fringes on the hind feet (below) trap air bubbles that support each stride.

Some shrews use a form of echolocation (as do bats, page 51) to detect obstacles or locate cover. Fast-moving and fast-living, most die before a year and a half.

Water Shrew (*Sorex palustris*)

These shrews seem more agile on the water than in it. One was seen dashing more than five feet across the surface of a pool, with head and body entirely out of the water. While they can also swim, their fur traps a film of air underwater, and they must paddle hard against its buoyancy. They seldom plunge in, preferring to forage among ground debris or in tunnels dug by mice and other little creatures.

Water shrews are usually found near running water. (The two above appear to size each other up on a creek bank in Colorado: one

grasshopper and two shrews—a sure formula for mayhem.)

They like wet, boggy areas in thick forests near free-flowing streams. Here they hunt insects, worms, spiders, and mollusks, and sometimes chase small fish.

Largest of the long-tailed shrews (genus *Sorex*) in eastern North

America, water shrews measure about 6 inches (152 mm), including a tail up to half that length. Breeding takes place between March and September. Females may bear three litters in a season, four to eight young in each litter. Since the mother has only six teats, probably only that number of babies will survive.

Well adapted to the cold, water shrews range from Nova Scotia across Canada to Alaska, and south in the western mountains to New Mexico, Arizona, and California. In the Appalachians they reach as far south as Tennessee.

Desert Shrew (*Notiosorex crawfordi*)

Prominent ears, silver-gray fur, and a short tail are trademarks of the desert shrew. These small, secretive animals range from the southwestern and south central United States into northern and central Mexico. Their range, like that of the pygmy shrew, is expanding.

They usually inhabit semidesert areas dotted with mesquite, agave, and scrub oak. There they may be found under boulders, beneath piles of brush, among low weeds near thorny scrub—and in woodrat nests. Although they will drink water when available, they don't need a permanent supply.

Some show up in beehives, where their size allows them to use the bees' own entrance. Adults measure only about 3.5 inches (90 mm) in total length.

As with all shrews, their eyes are tiny, and their vision—always less important than smell and touch—is poor. The desert shrew twitches its snout, "whiskering" its way around obstacles or in search of insects and other prey. To prevent the escape of crickets and cockroaches, a captive shrew first bit off their legs, then crushed their heads.

Reproduction extends through the warmer months. Nests of grass or leaves hold three to five offspring. Born blind and naked, at 40 days babies are near adult size, fully furred, and hunting on their own. Their only known predators throughout life are great horned owls and barn owls.

Pygmy Shrew (*Sorex hoyi*)

So diminutive is the pygmy shrew that its burrow in matted leaf mold could be taken for that of a large insect or earthworm. It can even convert a beetle's tunnel to its own use. Its overall length is between 2.6 and 3.8 inches (67 and 98 mm) with the tail taking up about a third. At an adult weight of 0.08 of an ounce (2.2 g), this tiny shrew weighs less than any other North American mammal. It has few rivals abroad too—a couple of European shrews and some featherweight tropical bats.

Pygmy shrews are fond of beech and maple or spruce and pine forests, where they make their homes in damp areas or dry places near water. They live under boulders, stumps, and rotting logs, or among leaf litter. Breeding habits are probably like those of the desert shrew.

Little is known about pygmies in the wild. In captivity they jump with agility, and can easily scale the sides of a cage and walk upside down on the wire top—sometimes hanging monkey fashion by their hind legs.

Biologists have studied some of these rare animals in the laboratory. They run so fast it is often hard to follow their movements. Tail held rigid, nose snuffing all around, they race about uttering shrill whispering and whistling sounds. When excited, they emit a musk so powerful it scents a whole room.

Their appetite seems especially incredible because of their size. In a 10-day period, one pygmy shrew ate its way through portions of 21 shrews, 20 houseflies, 3 mice, 2 crane flies, a beetle, and 22 grasshoppers.

Short-tailed Shrew (*Blarina brevicauda*)

"A ravening Beast. . .it biteth deep, and poysoneth deadly," warned a 17th-century clergyman. The short-tailed shrew is our only mammal definitely known to have a poisonous bite. Usually it has little effect on people, although one scientist, bitten on the hand, suffered shooting pains and severe swelling for days.

The shrew uses its weapon to subdue prey—sometimes larger mice or voles. The venom, mixed with saliva, seeps into bite wounds. But the short-tail feeds mainly on insects and other small invertebrates, and underground caches of live snails point to probable burrow "larders."

Relatively robust, this shrew spans about 5 inches (127 mm) from nose to tip of inch-long tail. Its thick fur is dark slate gray. The short-tail forages in woodlands, bogs, marshes, weedy fields, and anywhere else where surface vegetation provides cover. Here its runways form crooked lanes over the ground or wind beneath grass and leaves.

One of our commonest shrews, the short-tail ranges from Nova Scotia and New Brunswick south to Virginia, and west to Nebraska, the Dakotas, and Saskatchewan.

The SOUTHERN SHORT-TAILED SHREW (*B. carolinensis*), a similar species distinguished mainly by smaller size, is found throughout the corresponding southeastern half of the United States. Still another, *B. hylophaga,* ranges open country from Nebraska to Texas.

Family Talpidae

North America has seven species of moles. They live from southern Canada to northern Mexico, except in the deserts and high mountain ranges. Our sample shows three eastern moles. Western groups, while generally similar, include a shrew-size species. Lengths with tail range from 4 to 9.3 inches (100 to 237 mm) and weights from 0.3 to 6 ounces (9 to 170 g).

Moles spend most of their lives below ground, emerging only occasionally. Highly adapted for digging, they have powerful shoulder muscles and broad forefeet armed with long, flattened claws. Twisted palms face outward, the better to shovel aside the soil. Their cylindrical bodies, well designed for subterranean life, taper at both ends; and the velvety, dirt-repellent fur lies as well backward as forward, easing passage in any direction. Their ears—mere holes—have no outer flaps that would get in the way.

Moles' pinpoint eyes, buried in fur, can barely distinguish light from dark, but keen eyesight is no asset underground. Rather, the sensitive snout, tail, and facial whiskers scan surroundings by touch.

The coloring of North America's moles—black, gray, or brown—blends well with their earthy haunts.

Hairy-tailed Mole *(Parascalops breweri)*

A dense covering of bristly hairs on its tail distinguishes this mole from other eastern species. Its dark gray-black fur is coarse and often spotted with white on breast or stomach. Hairs on snout, tail, and feet turn pure white with age—three or four years maximum for all moles. Their natural predators—such as foxes, snakes, and owls—may shorten that span, but a greater risk is a flooded burrow.

Hairy-tailed moles favor the well-drained soil of high forests and fields up to a height of about 3,000 feet (914 m). This preference separates them ecologically from lowland-dwelling eastern moles.

The size of ridges and molehills formed by different species varies. Hairytails establish an elaborate network of tunnels, but there is scant evidence of them on the surface. Like other moles, they keep their tunnels in good repair.

Usually solitary, sexes mate in early spring. The female's vagina opens only just before her first breeding season—and closes again afterward until the next year. This prompted the old European folk belief that all moles were male until one year old, when half of them turned into females.

Gestation takes a month or so. In a nest about ten inches (25 cm) below ground and six inches

(15 cm) wide, four or five babies are born. Young hairytails shift for themselves within a month and breed in turn the following spring.

Eastern Mole *(Scalopus aquaticus)*

This large, common mole of the East measures about 7 inches (178 mm) long. It ranges from Massachusetts west to the Rockies and south to Texas and Florida. Like all moles, the eastern mole

Clawing the earth loose with one forefoot, the mole shoves it upward, heaving up the turf. These tunnels are feeding and resting zones, well supplied with insects and earthworms.

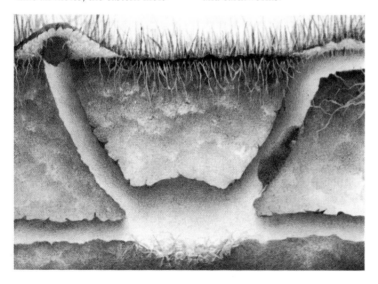

can swim, though it does so less than its Latin name implies. It lives in moist, sandy, or loamy soil in woodlands, meadows, pastures, gardens, and lawns. There it digs a maze of burrows (above).

Moles make two kinds of tunnels. In warm, damp weather they build shallow subways marked by their telltale surface ridges.

When the weather is dry or cold, the mole burrows deeper, pushing surplus dirt out to form the mounds we call molehills. The deep tunnels are its proper home, where it builds a nest—lined with grass, rootlets, or leaves—and females raise their young.

Star-nosed Mole *(Condylura cristata)*

In appearance this is certainly one of our most peculiar animals. Its slender nose ends in a crudely star-shaped growth—a naked disk surrounded by 22 symmetrically arranged pink, fleshy tentacles. These feelers wave around all the time as the mole forages.

The starnose is smaller than the eastern mole, with less exaggerated forefeet and a much longer tail. During late winter and early spring, its tail thickens to the size of a pencil with stored fat—probably a food reserve for the breeding season.

Unlike other moles, starnoses live in small colonies, and couples pair off in the fall, remaining together until the young are born in the spring. The babies' stars are clearly visible at birth.

This mole is less subterranean than most, and makes runways on top of the ground or snow as well as beneath. It leads an amphibious existence, often tunneling into the banks of streams a foot or so underwater.

An excellent diver and swimmer, it paddles with its forefeet and uses

its tail as a rudder. In winter the starnose roams the ice or swims below it, in search of the aquatic worms, insects, and crustaceans that make up three-quarters of its diet, the rest being land animals.

Bats

ORDER Chiroptera

Sonar on the wing: Ultrasonic cries of the gray bat pulse from the open mouth as the sensitive ears listen for echoes. Legs and "hand-wings" working together, a bat swims through the air.

ONE-FIFTH OF ALL mammal species are bats. To the writer D. H. Lawrence, bats had "wings like bits of umbrella." For the Chinese they symbolize good luck and long life, and appear in art as emblems of health and virtue. Though bats exist almost everywhere, their secretive behavior and nocturnal lifestyles keep them out of sight. Most of us are unfamiliar with bats, even those that nightly dart through twilight skies over our homes. During a recent trip to Guaymas, Mexico, on the Gulf of California, my inquiries regarding fish-catching bats typically were met with incredulity. The residents had never heard of any such bats and clearly doubted their existence. I soon learned why.

To reach my first Mexican fishing bat roost, I went out in a small boat with my colleague and wife, Diane Stevenson. I had to swim to a tiny cliff-faced island guarded by breakers that threatened to dash me onto sharp, barnacle-encrusted rocks.

Then, after climbing some 40 feet (12 m) up a slippery, guano-covered cliff, I was still 10 feet (3.5 m) short of a chattering colony of perhaps a thousand bats. Deep in dark crevices, they were not even visible. In the sea again several minutes later, I treaded water, waiting to be picked up. My bleeding hands—and the realization that a nearby village was devoted to shark fishing—conjured up visions of hungry-eyed sea monsters lurking beneath.

Only a few researchers have ever seen a Mexican fishing bat alive. Yet, with luck, Diane and I hoped to capture and photograph some of these fascinating animals. We had equipped ourselves with a powerful light, a night-vision scope (which amplifies available light about 64,000 times), and special, fine-threaded mist nets. Still, on a lengthy drive along the coast that night, our spotlight revealed only two fishing bats. With the scope

we could see that the bats were restricting themselves to a well-defined area of shoreline about 100 yards (91 m) long. This suggested that fishing bats, like many other kinds of bats, probably return nightly to their customary hunting territories.

On the following night we waited on the beach, hoping. Our mist net—40 feet (12 m) long and 7 feet (2.1 m) high—was now strung across the bay exactly where the bats had fished. Suddenly our net poles shook violently, and nearly half the net disappeared into the water beneath. To our dismay we found that a two-foot Mexican needlefish had jumped into the net and rapidly was destroying both our net and any chance of catching bats that night. Then, just as all seemed lost, seven fishing bats struck the net. Frantic, we fought to disentangle a large fish and seven bats all at once. Our unexpected good fortune elated us. Sound from the struggling fish apparently had lured the curious bats into our net.

Clearly, there is good reason to be ignorant about bats when even bat researchers sometimes find them elusive. However, years of effort by many researchers have disclosed an impressive array of adaptations of great interest to humans. Bats are likable creatures when we take the time to get acquainted. They are highly sophisticated, more closely related to us than to the mice with which they are often compared. Both bats and people evolved from similar insect-eating ancestors, and, as early as the Eocene, about 50 million years ago, bats were clearly bats.

Chiroptera—the name means "hand-wing"—are the only truly flying mammals. The bones of the wing are the same bones as in the human arm and hand, but the bat's hand expands to form its highly maneuverable wing. Between the fingers and between the arm and the body and the outside of the leg stretches a double membrane of skin. In many species, another extension of skin (called the interfemoral membrane) reaches from the heel of the hind limb to the tail. A bat does not glide like a so-called flying squirrel but, like a bird, propels itself, moving its wings in a figure eight or an ellipse.

Bats are divided into 18 families and almost 1,000 species—second in number only to the rodents. Some 100 bat species live north of the southern Mexican border. Worldwide, sizes

range from the South Pacific's flying fox, with a wingspan of more than five feet and a weight of two pounds or more, to the Philippine bamboo bat, which weighs less than a dime. Bats are long-lived. In contrast to other small mammals, some, such as the little brown bat, may live 30 years or more.

Members of even a single North American family, the Vespertilionidae, exhibit a striking range of color, size, shape, and behavior. Though most of the family members feed only

upon insects, pallid bats also include scorpions in their diet. The Mexican fishing bat is in this family. Members of the family Phyllostomidae, American leaf-nosed bats live mainly in the tropics, but a few extend their ranges into northern Mexico and the southwestern United States. These include fruit-eating bats, a variety of nectar-feeding species, some insect eaters, and the vampires—which do not reach the United States.

Leaf-chinned bats of the family Mormoopidae are closely related to the leaf-nosed bats and also prefer tropical climates. Like the following families, they eat only insects. Funnel-eared bats of the family Natalidae are delicate and brilliantly colored, while the less showy members of the Molossidae have impressive adaptations for crowded living and long-distance flight.

Mexican free-tailed bats, for instance, form the largest colonies of any mammal. Some Texas cave colonies contain as many as 20 million individuals. Each evening, as virtual clouds of Mexican freetails emerge from many caves in the southwestern United States and Mexico, one can hear the muffled roar of the simultaneous fluttering of tens of thousands of bat wings.

The largest colonies require more than 100,000 pounds (45,455 kg) of insects nightly, and to obtain them, bats may cover more than 100 miles (160 km) in a single night. Great flocks sometimes climb 10,000 feet (3,000 m) above ground, where airstreams are believed to aid them in attaining speeds as great as 60 miles (96 km) an hour. These jetsters of the bat world have long, narrow wings designed for speed and endurance. One morning I watched large flocks spiral down from high in the sky. During reentry into their vertical cave opening, many dived with swept-back wings at speeds above 80 miles (128 km) an hour.

In contrast to these colonial cave dwellers, hoary and red bats—of the family Vespertilionidae—live alone in trees, where their unusual color patterns blend with the foliage or bark and protect them from discovery. Like the Mexican freetails, they

Beads of condensed moisture glisten on the fur of an eastern pipistrelle in hibernation. Its body temperature—48° to 52°F (8° to 11°C)—is about the same as the deep recess of the Kentucky cave where it hangs, frequently motionless for weeks at a time. This species probably spends more time in hibernation than any other North American bat. In Wisconsin they may hibernate from late August into May.

Loosely spaced on a vertical wall, hibernating gray bats (opposite) dissipate body heat and maintain a constant low temperature. About two-thirds of the known gray bat population winters in this northern Alabama cave. In the spring these bats will fan out, in colonies, to summer roosts in nearby states (map on page 61). But each fall they will return, loyal to this cave for life.

migrate. In autumn, hoary and red bats fly south from Canada and the northern United States. Some may reach the Gulf states, where they roost in relatively exposed places. They pass into torpor in cold weather but rouse themselves and feed when warm spells permit insect activity. Such winter feeding often takes place in broad afternoon light.

Eastern pipistrelles and most *Myotis* species, on the other hand, retreat to caves for the winter. There they hibernate six months

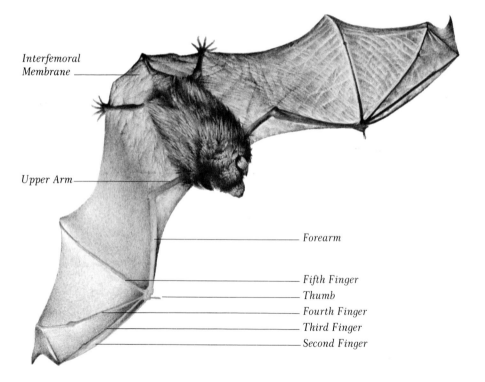

Interfemoral
Membrane

Upper Arm

Forearm

Fifth Finger
Thumb
Fourth Finger
Third Finger
Second Finger

or longer without food. To accomplish this, some may double their lean body weight with fat accumulations during the late summer.

Some bats actually travel north to winter. Most gray bats living in northwestern Florida migrate more than 300 miles (482 km) north to a single cave in northern Alabama. This rare cave traps cold air all winter but does not freeze beyond its entrance area. As the bats arrive in September, they mate. Females enter hibernation quickly, storing sperm in the

uterus all winter. Ovulation and fertilization follow in the spring, when the bats become active again. Hibernation at stable low temperatures seems necessary to delay pregnancy, since, if a female bat is taken from hibernation and kept active in a laboratory, she will become pregnant, even in midwinter.

Each winter this Alabama cave houses the world's largest known hibernating bat population. In 1969, when the cave was discovered, it held 1.5 million gray bats. In places, a solid mat of bats covered the cave walls as far as one could see. I estimated the bats' weight to exceed 16 tons (16,300 kg).

The summer roosts are in warmer caves near rivers or lakes. Mothers may fly over the water for more than seven hours in a night, catching insects in the air or on the water's surface before returning to nurse the young. A single gray bat may eat more than 3,000 insects in one night and, during times of maximum insect abundance, may catch two or more insects in as short a span as five seconds. Digestion is rapid, and in-flight defecation permits continuous feeding. Assimilating more energy than she expends in flight, the bat apparently puts the energy to use, for the breasts of mother bats returning to the cave are distended with milk.

To study these bats, Diane and I marked them with bands coded with reflector tape. With our night-vision equipment we could recognize individuals. We showed that some females defend the same feeding territories in successive years, and that they sometimes permit other females of the same species to share these while aggressively chasing away all others. They apparently recognize each other through vocal communication.

The use of sound reaches a pinnacle of sophistication in the order Chiroptera. Many species navigate, locate and avoid obstacles, feed—all by sonar signals. The bat produces sound with its larynx, just as people do, by causing air to vibrate. The faster the vibration, the higher the pitch of the sound. In some bats, a pulse of sound is made up of vibrations of a constant frequency—a single note. But in others, the pulses shift frequency from high to low. The human ear can hear sounds that range from about 20 vibrations, or cycles, per second, to about 20,000. Most bat cries are of higher frequencies and not audible to us. The little brown bat, for instance, emits sounds that often sweep from 90,000 cycles per second down to about 45,000. However, some bats, such as the western mastiff and the spotted bat, emit cries that we *can* hear even though the mastiff may fly 1,000 feet (305 m) above our heads.

Such sonar is called echolocation because the bat produces a short pulse of sound and listens for the echo to bounce back from nearby objects. The difference between the time of emission of sound and the return of the echo measures the target distance. Variation in echo intensity can gauge direction. Because an object's shape and texture reflect some frequencies more strongly than others, the cruising bat can sift information from the varying echoes and distinguish between an obstacle and edible prey.

When an insectivorous bat detects an insect, the bursts of sound increase from cruising levels of 5 to 20 pulses per second to buzzes of as many as 200 pulses per second. This, plus additional changes in frequency modulation, permits the bat to read the exact location and nature of its meal. In the laboratory, bats can distinguish between mealworms tossed in the air and other objects of similar size and echo strength. Many bats are able to choose favorite sizes and species among available insects.

Some kinds of moths, however, enhance their chances of survival by listening (by means of ear cavities on the sides of their thorax) for bat sonar and taking evasive action. Some respond with their own high-frequency signals that seem to discourage the bats. Or, to evade the predator, they soar upward, loop, or plummet toward the ground.

There is a mite that parasitizes moth ears. A moth deaf in both ears is more likely to be eaten by a bat, thereby also ending the mite's life. The first mite to find the moth leaves what is believed to be a chemical trail that leads to the ear it parasitizes. Following mites use that first trail and thus do not destroy both of the moth's ears.

Bat sonar is many times more efficient, weight for weight, than any sonar or radar developed by humans. Even though thousands of animals may be flying together through a cave—in what has been called "a veritable ultrasonic Babel"—each bat can follow its own signals. The bats also discriminate

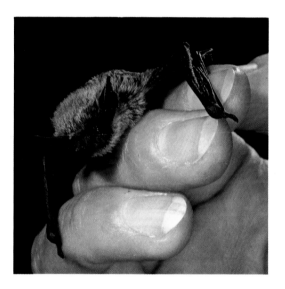

information-carrying echoes from background noise. They detect prey skillfully at ground level—something radar systems have great difficulty doing. Many researchers believe that an understanding of the bat's navigation and sonar will lead to insights into the nervous system of other mammals.

With sonar, bats catch prey smaller than mosquitoes, even on the darkest night. Bats also eat many other harmful insects, such as cutworm moths, corn borer moths, grasshoppers, and locusts. Little brown bats commonly live in colonies of a dozen to several thousand. A colony of only 500 easily can capture 500,000 insects in a night—a clearly beneficial element in nature's balance.

The noninsectivorous species also make

North America's smallest bat, the western pipistrelle, here weighs in at 0.14 of an ounce (4 g). Its body as pale as the desert canyons where it often roosts in rock crevices, this voracious midget flutters out early in the evening to eat perhaps one-third its weight in insects, naps until dawn, then hunts again.

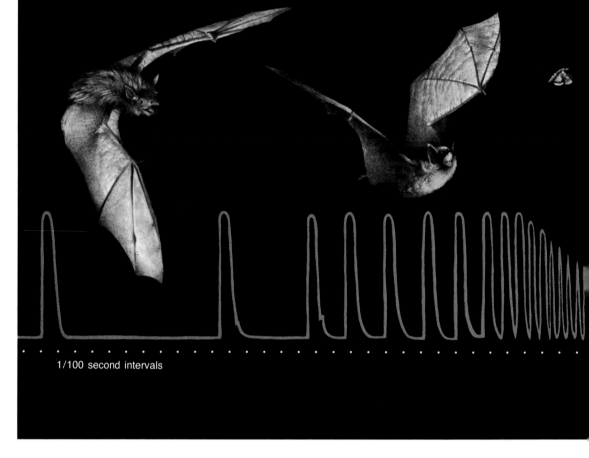

On the hunt, a little brown bat finds a moth, scoops it up in a handy wing— from the tip it will be popped into the bat's mouth—and flies on to pursue another morsel. Sound signals pulse from the bat's open mouth, 10 clicks a second, as the bat cruises, avoiding obstacles and searching for prey (left side of graph). When the bat hears an echo that translates "food," the sonar quickens to a buzzing 200 clicks a second. The higher frequency brings precision as the bat homes in on its target (center of graph). At the point of catch, signals stop. While the bat eats, it flies in silence (shown by the smooth line), but less than one-fifth of a second later (the single peak at right) the supersonar resumes.

1/100 second intervals

their contribution to human welfare. In the tropics many trees and shrubs depend on nectar-feeding and fruit-eating bats for pollination and seed dispersal. Several economically important fruits—avocados, mangoes, breadfruit, guavas, and probably bananas—come from plants that are believed to have been originally dependent upon bats. All North American bats possess sonar but the chiefly noninsectivorous species probably use sonar primarily for navigation and rely more on their excellent smell and vision when hunting food—for bats are not blind.

Unfortunately, too few people know the many fascinations of these creatures, and instead hold deeply ingrained fears and prejudices based upon myths. All bats *can* see, some only less well than others. Bats never really become tangled in human hair. They are shy and avoid people whenever possible. They are neither verminous nor filthy. Their parasites certainly are no worse than those of dogs, cats, or other mammals, and they groom themselves thoroughly.

Bats are not serious threats to human health, as often feared. In fact, only 23 cases of bat-transmitted human rabies have been recorded in the United States since 1946, when record keeping began. Far more people die from dog attacks, bee stings, power mower accidents, or even from being struck by lightning. Humans rarely are bitten by bats except during careless handling of sick individuals.

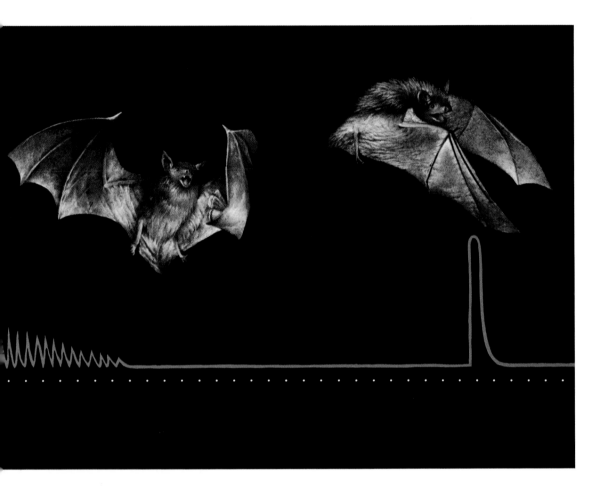

Despite constant work with bats worldwide, not one of hundreds of bat researchers is known ever to have been attacked. I have come into close contact with millions of bats of some 150 species in a dozen countries over a period of 20 years. Repeatedly I have been impressed by their gentleness.

Numbers of several North American species have declined dramatically in recent years. Some, such as the gray bat, are now endangered. In some areas, populations of even the Mexican free-tailed bat have dropped to but one percent of their former size. The overuse of chemical pesticides has both decreased the food supply for insect-feeding bats and—since many bats must ingest the poisons as they eat—polluted it.

Every year thousands of bats die when people disturb maternity or hibernating roosts. In winter, the presence of cavers repeatedly arouses bats and depletes their limited fat reserves. The bats may starve before spring renews their food supply. Most bats that are frightened into abandoning their chosen hibernation sites probably die. If gray bats are disturbed continually, rather than switch to another cave, they generally keep returning until exhaustion brings death. Bats that do not dwell in caves often are killed because they live near people who needlessly fear them.

As humanity's demonstrated allies, bats deserve all the respect and consideration we can give them. Their decline certainly is not in our best interest.   MERLIN D. TUTTLE

Family Mormoopidae

Folds of skin and bristly hairs around the mouth distinguish the eight species of this family. The skin distends to form a funnel of the mouth, while the hairs appear to direct airflow inward in a manner that may enhance the bat's ability to feed on insects, its sole food. One colony of half a million mormoopids observed in Mexico consumed about 3 tons (1,000 kg) of insects nightly.

Small in size—they weigh between 0.25 and 0.63 of an ounce (7 to 18 g)—these bats vary in color from brown or gray to bright orange. Half of their tail protrudes beyond the tail membrane. Mormoopids occur from southern Arizona and Texas to Brazil. Mainly cave dwellers, they often form sizable colonies—to our benefit. Farmers use for fertilizer the large amounts of guano produced by the roosting bats.

Leaf-chinned Bat (*Mormoops megalophylla*)

This bat, distinguished by the fold of skin across its chin, is the only mormoopid that lives in the United States. Often skimming no higher than 6 feet (1.8 m) over land or water, *M. megalophylla* uses its mouth as a megaphone to beam pulses of sound through the dark. Once the quarry is detected by the signal's rebound, the bat veers to catch the insect.

The leaf-chinned bat has fur that ranges in color from reddish brown to buff. It occurs from Arizona and Texas south to northern South America and the offshore islands of Venezuela. It shelters in habitats that vary from desert scrub to tropical forest. This bat colonizes in caves—sometimes with large groups of other bat species—and in mines, tunnels, and an occasional building.

Not much is known of this insect eater. A confirmed nomadic streak in the species impedes research. A colony may stake out several roosts over a 200-mile (322-km) range. One Mexican cave harbored a population of 500,000 one November. Two months later, the cave was deserted.

Family Phyllostomidae

The bat family most diverse in numbers of genera, American leaf-nosed bats include about 150 species and range from the American Southwest to northern Argentina and the West Indies. Thirty-two species live in North America, north of Mexico's southern border. Wingspans vary from nine inches (229 mm) to the three-foot (914-mm) spread of the false vampire bat, the largest in the Western Hemisphere. Phyllostomids roost almost anywhere. They eat insects, fruit, nectar (below), frogs, lizards—even small birds and mammals. They serve as important dispersers of seeds and pollinators of tropical night-flowering plants.

A protrusion from the snout known as a "leaf" usually distinguishes these bats. Through their nostrils most emit a weak sonar signal. As a result, the many species of small phyllostomids are nicknamed "whispering bats."

Southern Long-nosed Bat (*Leptonycteris curasoae*)

Southern long-nosed bats live in the desert scrub country of Arizona, New Mexico, and Mexico. Northern populations apparently migrate to Mexico in winter. A gray to reddish-brown bat, *L. curasoae* has no tail. It inhabits caves and mines, where it hangs from the ceiling with its feet so close together it can twist nearly 360° to regard an intruder.

The blossoms of agave (left), giant saguaro (above), and other desert plants attract this bat. The 0.63- to 1.06-ounce (18- to 30-g) mammal often appears to hover, hummingbird-like, to insert its long tongue in a flower. Fleshy bristles on the tongue increase its effective surface area and permit the bat to brush up large amounts of nectar, high in sugar readily convertible to energy.

Glands secrete hydrochloric acid to break down the ingested pollen proteins into amino acids—needed for the building of body tissue. After the flowering season these bats eat fruit.

They feed for several hours during the night, repaying the debt by pollinating plants in their travels between blossoms, in a symbiotic relationship known as chiropterophily—bat-love. Some 200 genera of bat-loving plants, with large amounts of high-calorie nectar, depend on bats for survival, and vice versa.

Today the agave plants of northern Mexico are harvested intensively to make tequila. Many large bat colonies that depended on these agaves have disappeared.

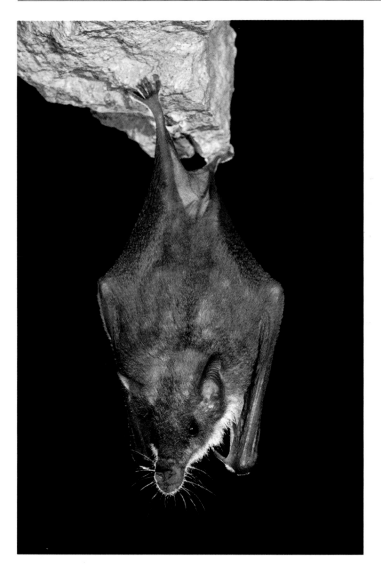

Hog-nosed Bat (*Choeronycteris mexicana*)

The absence of lower incisors facilitates tongue movement in the hognose. That trait and a long, slender nose adapt this nectar feeder to foraging in the flowers of cacti and other desert plants. In the United States, the hognose occurs only in the canyons of the far Southwest. From there it ranges south through Mexico and Central America to Honduras. This bat weighs between 0.35 and 0.71 of an ounce (10 to 20 g). It lives near dimly lit cave entrances. Like the other phyllostomids, *C. mexicana* has weak sonar. Some insectivorous bats have a signal up to 1,000 times stronger. When threatened, it chooses to fly from the cave rather than retreat, utilizing the comparatively good eyesight of leaf-nosed bats.

Adult females carry embryos equal to as much as a third of the mother's weight. The newborn are already well furred, unusual in North American bats. The hognose inhabits mainly arid country at elevations of about 2,000 to 8,000 feet (610 to 2,440 m).

California Leaf-nosed Bat (*Macrotus californicus*)

Though barely able to walk on the ground, *M. californicus* more than compensates in the air. Short, broad wings give it maneuverability. Large ears enhance its hearing and enable this foliage gleaner to discern insects amid the thick vegetation in which it often hunts. Seeming to hover over trees and other plants, it forages for nonflying, less mobile prey, such as caterpillars and grasshoppers.

This bat, which weighs little over half an ounce (14 g), ranges from southern California, Nevada, and Arizona into Baja California and Sonora, Mexico.

California leaf-nosed bats seek roosts in mines and large caves, but they do not cluster. When individuals touch each other, they grow restless. Often they take walks across the ceiling, swinging from one leg to the other and hanging on to the horizontal surface with their large feet and long nails.

Hairy Fruit-eating Bat *(Artibeus hirsutus)*

Of our fruit-eating bats, this one lives farthest north. It is found over a 600-mile (965-km) strip of semiarid land running along Mexico's coast from Sonora to Guerrero.

Related species of *Artibeus* extend the range of the genus south through Central America into South America and east to the West Indies. These bats do not hibernate; they live in climates warm enough to produce food year-round.

A blunt, round face striped—sometimes faintly—with white, a nose leaf, and the absence of a tail mark the genus. An adult *A. hirsutus* measures from 3.1 to 3.4 inches (79 to 86 mm).

Many fruit-eating bats appear specialized in their preference for figs. *A. hirsutus* often carries a fig to a temporary roost before eating (above), for a thorny palm tree offers protection from predators. The bat crushes the fig between its jaws to extract the juice, and spits out the pulp.

A. hirsutus forms small cave colonies, but for several related species the roost often takes the form of a "tent," which the bat fashions by biting palm fronds so they will fold over. By roosting away from food sources, the bats avoid the opossums, owls, and snakes that wait in the fruit trees seeking a quick meal in the form of a bat.

Common Vampire Bat (*Desmodus rotundus*)

Of the more than 100 bat species north of Mexico's southern border, only *D. rotundus* is an economic threat to humans.

Feeding solely on fresh blood, the common vampire usually preys on sleeping animals, often livestock. This vampire alights, selects an exposed place on the skin, and makes a painless incision. (All three vampire species have sharp teeth.) The bat then forms a tube with its tongue and draws blood. Anticoagulants in the saliva and the abrasive, darting tongue assure a continued flow.

These one-ounce (28 g) mammals may guzzle over half their weight at a sitting yet remain nimble, thanks to the strong hind legs on which they can hop away when an 1,100-pound (500-kg) cow flicks its tail or rolls over. Rapid urination decreases the weight of the meal they must carry off. Blood loss seldom threatens cattle. But some vampires carry diseases, and open wounds invite parasites and infection. Damage to livestock from northern Mexico to Argentina and Chile—the range of the vampire—is said to amount to millions of dollars a year.

Common vampires colonize, usually in groups of about 100. Even in their caves, hollow trees, old wells, or abandoned buildings, these are acrobats. They can walk or tiptoe, leap or scramble or fly. If disturbed in a roost they quickly retreat into crevices. When *D. rotundus* leaves its roost after dark, it flies slowly, silently, about three feet from the ground.

The razor-edged incisors bared by a juvenile (top) and the swift tongue of an adult (above) are the dining implements of D. rotundus.

Family Natalidae

A single genus and five species compose this little-known family of insectivorous bats. Natalids live primarily in tropical lowlands, but they also inhabit caves and mines as high as 3,000 feet (914 km) above sea level. Their range extends from northern Mexico to the West Indies and central Brazil.

Prominent funnel-shaped ears give the family its popular name. Sticklike legs, sometimes longer than the small head and body, help bestow a delicate appearance. The genus name, *Natalus,* comes from Latin, "related to one's birth," because even as adults these bats have a newborn look about them. Less than 0.35 of an ounce (10 g) in weight and often brightly colored, these small bats flit through the air, looking more like moths than mammals.

Greater Funnel-eared Bat (*Natalus stramineus*)

Small eyes and large ears offer clues to this bat's behavior. Guided by hearing, this species beams a high-frequency sonar signal. The signal, sweeping up to 85,000 cycles a second, finds tiny insects—prey that more languorous sound waves perhaps would bypass. The bat's ear, its large outer surface studded with numerous glandular papillae, serves as a keen receiver.

N. *stramineus* weighs about 0.18 of an ounce (5 g) and measures up to 2 inches (51 mm) in total length. It occurs from northern Mexico through Central America to Brazil, living in the darkest recesses of caves, where it may congregate with many—or with as few as a dozen—of its own species.

Gray after molt, these bats often turn brilliant orange or yellow, bleached by the ammonia fumes that the accumulations of guano produce in their caves.

Family Vespertilionidae

This large, varied family encompasses about 40 genera and some 350 species worldwide. Their muzzles are unornamented by a nose leaf or a chin flap, and their long tails are almost enclosed within the interfemoral membrane. Beyond these, common qualities are few. Vespertilionids weigh from 0.12 of an ounce (3.4 g) to 2.82 ounces (80 g).

Though most species eat insects, some forage on the ground while others hunt on the wing. Almost all send ultrasonic echolocation signals through the mouth. Populations vary in character. Gray bats, for instance, rely on colonial behavior for survival, while some tree dwellers live secret, solitary lives.

Members of this family occur in all but Arctic regions and have adapted to almost every kind of habitat. The genus *Myotis* is more widely distributed than any other mammalian genus except possibly *Homo*— to which we belong.

Little Brown Bat (*Myotis lucifugus*)

One of the most abundant bat species found in North America, M. *lucifugus* ranges from California and Alaska east to Newfoundland and as far south as Georgia. The little brown bat weighs about one-quarter of an ounce (7 g), and is cloaked in dense, glossy fur—tan, reddish brown, or dark brown.

Little brown bats mate in the fall in caves where they hibernate. Fertilization occurs with their emergence in spring; gestation takes 50 to 60 days. Females then form maternity colonies—which number from as few as a dozen to thousands—in very warm places. Probably, M. *lucifugus* originally roosted under old bark or in hollow trees. But such trees are now in short supply and buildings often substitute. A colony was found in an attic with the temperature close to 130°F (54°C).

Single males also roost in buildings but sometimes shelter in rock crevices or beneath shingles, siding, or bark.

At some time from late May to early July, each mother bears a single infant. She receives it in the interfemoral membrane, cupped like a basket. The baby aids in its own delivery by pulling its way out of the womb. Attaining adult weight within a month, the young bat must quickly fine-tune its insect-seeking sonar in order to store enough body fat for the winter ahead.

One study has found these bats specializing in mosquitoes one season and mayflies another. The diet also includes flies, moths, and beetles.

Keen's Bat *(Myotis keenii)*

Similar in size and color to the little brown bat, *M. keenii* is distinguished by the longer ears that suggest it is a foliage gleaner, a bat that often garners insects from vegetation rather than catching them on the wing. This bat's large ears may adapt it to the relatively more complex physical environment in which it hunts.

M. keenii is more solitary than most other *Myotis* species. Entering caves in winter, it usually remains alone, rarely hibernating in groups. Less than 0.4 of an ounce (11 g) in weight, it squeezes into rock crevices and easily can be overlooked. In spring it emerges and virtually disappears from sight. Individuals and small colonies occasionally have been discovered roosting beneath the bark of old trees. Possibly this is where they raise their young; there is some evidence that pregnant females prefer heavily forested areas. But no one knows for sure.

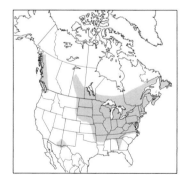

Gray Bat *(Myotis grisescens)*

This quarter-ounce to half-ounce (7- to 14-g) bat lives in the southeastern United States. It colonizes caves that usually are within a mile of bodies of water with large populations of prey: mayflies, other aquatic insects, beetles, moths. (Above, the meal is a rosy maple moth.) *M. grisescens* is distinguished in its genus by the uniform color of each hair.

Foraging for food expends a great deal of energy, particularly between May and July, when mothers rear their young. A mother must transfer enough energy, through her milk, to maintain the young's temperature at about 102°F (39°C)—more than twice the cave's temperature.

In late summer the bats home to their winter caves. To facilitate torpor, they must find cold but not freezing roosts. Migrating as far as 300 miles (483 km) one way, 95 percent of the known population hibernates in only 9 caves. Thus concentrated, they are extremely vulnerable to human disturbance.

Censuses once showed serious population decline, but certain protected colonies are improving.

The map shows migrations of summer colonies to a single winter site in Alabama.

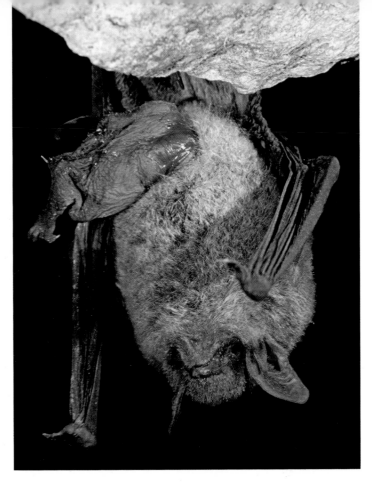

Week-old gray bats cluster for warmth in an eastern Tennessee cave. Nurseries such as this may contain 300 young per square foot (929 sq cm), yet mothers apparently identify their own offspring.

Bearing a single young (top), the mother attaches the featherweight, hairless baby to the ceiling, where its tenacious grasp is instinctual (lower). If the infant falls, it likely will die. Within three weeks the young bat must fly, begin to hunt with its sonar, and, returning to the cave, execute a midair flip and grasp the ceiling. Only about half survive to become adults.

Silver-haired Bat (Lasionycteris noctivagans)

A striking pelage of black tipped with silver marks this bat. Often living alone, deep in the forests of the northern United States and Canada, the silver-haired bat roosts under bark and leaves and in holes made by woodpeckers. It emerges late in the afternoon and usually seeks a nearby pond or stream. There it hunts for insects, looping around and around the same circuit. The silver-haired bat's flight, at about 10 miles (16 km) an hour or less, seems leisurely compared to that of other species. The bat is thus well named—noctivagans is the Latin for "night wandering." With the possible exception of the western pipistrelle, the silver-haired bat may well be the slowest flier of all North American bats.

L. noctivagans, which weighs less than 0.35 of an ounce (10 g), is an inveterate wanderer when it comes to range. In winter this species hibernates in cliff faces, hollow trees, and buildings across the continent. In inland areas it winters in the southern states. Along both coasts, where the climate is milder, it hibernates from the southern states at least as far north as New York and Oregon. But migrating individuals have turned up in such disparate places as Maine's coastal islands and Bermuda—about 650 miles (1,045 km) from the North American mainland.

Eastern Pipistrelle (Pipistrellus subflavus)

A sprite that weighs as little as 0.14 of an ounce (4 g), this bat is recognized by its distinctive tricolor fur. The hair is dark at the base, light in the middle, then dark again at the tips. The eastern pipistrelle, a woodland dweller, occurs from Nova Scotia west to Minnesota and south through eastern Mexico to Central America. It is perhaps the most common bat in eastern North America. Some females mate in the fall and store the sperm over the winter; others mate in the spring. All give birth to one or two young by early summer within colonies high in caves, cliffs, or buildings.

The pipistrelle's summer and winter ranges seem to coincide, as recoveries of banded individuals seldom show movement of more than 50 miles (80 km). In warm months the bats sometimes roost in dense foliage. At dusk they flutter through trees and over streams in search of insects—bats so small they themselves can be mistaken for large moths.

Pipistrelles hibernate in more caves of eastern North America than any other species, with some individuals occupying the same spots in the same cave over successive winters. P. subflavus can discern between places in the cave where the temperature might differ by only a few degrees, thus even by a little increasing its chances for survival to spring.

Big Brown Bat (Eptesicus fuscus)

Size and color distinguish the big brown bat, which often weighs up to three-quarters of an ounce (21 g). Its fur, black at the roots, lightens to brown on the surface. Found from Canada to northern South America, this is second only to the hoary bat in its ranging of the Western Hemisphere. Perhaps its wide distribution derives in part from the way it adapts to the human world. Big brown bats often roost in barns, belfries, and attics. Not surprisingly, more than any other species they come into contact with people.

E. fuscus circles over and over the same territory. It has a remarkable homing ability. Some individuals were released 250 miles (402 km) from their roosts. They returned, averaging 50 miles (80 km) each night.

This bat forages for food in a variety of habitats. Adults can eat their fill of insects in an hour. They often gain a third or more of their lean weight in fat before entering hibernation. Some have been observed awake in sub-zero temperatures. In northern areas, whole colonies of these hardy bats hibernate with body temperatures as low as 29°F (-1.6°C). Some big brown bats, perfectly healthy, have been found partly covered with ice.

Mexican Fishing Bat (Myotis vivesi)

Huge feet and claws that look like pitchfork tines identify this bat, one of two Mexican species that fish for a living. Found on the coast and islands of the Gulf of California. M. vivesi often roosts in the crevices of cliffs rising from the sea. On some remote islands, these bats simply live in spaces beneath rocks on the ground or even in old turtle shells. When darkness falls, the fishing bat swoops low over the water. It locates fish through sonar that echoes from their tiny dorsal fins or from the surface ripples. The bat then drags its claws through the water, gaffing such prey as minnows and crustaceans. M. vivesi weighs from 0.63 to 0.81 of an ounce (18 to 23 g). It is very similar to other species of the genus Myotis save for its formidable feet and claws.

OVERLEAF: A Mexican fishing bat carries off its catch—a mullet speared with quarter-inch-long (0.6-cm) claws, then shifted to its mouth. Unlike most bats, M. vivesi is an adept walker.

Red Bat *(Lasiurus borealis)*

A furry interfemoral membrane, long wings, short ears, and color distinguish the red bat, whose fur ranges from bright orange to light rusty brown. It is one of the few species in which male (above) and female show a marked difference in color. Usually males are redder, females more frosted with white. It weighs about 0.25 to 0.42 of an ounce (7 to 12 g). Members of this genus are among the few species that bear two or more young at each birth.

L. borealis is secretive, given to roosting alone in dense foliage, where it hangs limply by one leg. Its resemblance to a dead leaf may fool a preying blue jay, opossum, hawk, or owl.

This bat usually hunts insects within a mile of its roost—often around streetlights in towns. The seeming loyalty to a particular foraging site suggests that the species is territorial.

L. borealis is found throughout North America, except for the extreme north, the southern half of Florida, and arid regions. The red bat migrates south within its range in winter. It copes with subfreezing temperatures by varying its heart rate and wrapping its furred interfemoral membrane around its body like a blanket.

Hoary Bat (*Lasiurus cinereus*)

White-tipped brown fur gives the hoary bat a dazzling, frosted appearance—and its name. While the red bat roosts mainly in deciduous trees, the hoary prefers evergreens, where its silvery coat, blending with the lichens that grow on tree trunks, may conceal the bat from a hungry hawk or owl. Like all bats, it grooms itself thoroughly (above), licking its fur as a cat does.

L. cinereus ranges from Canada to Argentina and Chile and has been seen in such far-flung places as Iceland and the Orkney Islands. It is the only wild land mammal to have reached and established itself in the Hawaiian Islands. North America's most widespread bat, it is more common from the prairie states west than it is in the East. Even so, this tree bat is usually seen by people only after a summer storm. That is when a female, burdened by clinging young and wet fur, might be shaken from a tree like ripe fruit.

The hoary bat is primarily insectivorous, and—at up to one and one-quarter ounces (35 g)— large for its genus. Long, powerful wings permit the long-distance travel and rapid flight—during which these bats often chatter audibly. Spring and fall, they migrate in great waves. In spring the females, already pregnant, move north ahead of the males. They do not form colonies. One Wisconsin hoary bat returned for at least three years to the same spot on a blue spruce branch to rear her young.

Spotted Bat (*Euderma maculatum*)

Oversize pink ears and black fur symmetrically splotched with white make the "elusive and beautiful" spotted bat a standout among the species of North America. A medium-size bat, *E. maculatum* has the largest ears of any bat on the continent—they measure nearly two inches (51 mm) and adapt the bat for foraging among the rocky outcrops and scrub of the western United States and Mexico. In addition, the large surface of the ears may play a thermoregulatory role in helping the bat dissipate body heat in the desert climate.

Until recently this reclusive animal was thought to be rarer than it actually is. It was not discovered until 1890, and even now is not well known. Highly specific needs in habitat doubtless limit its numbers. It apparently roosts alone in crevices in high cliffs of sedimentary rock. Individuals also maintain exclusive territories in which they forage swiftly, well above the ground.

This bat can walk across flat surfaces, apparently with ease, by using its feet and wrists. It has been seen pursuing prey—possibly grasshoppers or beetles—in this manner. However, the staple of its diet seems to be moths. Researchers believe the spotted bat, following the habit of other bats, removes the head, wings, and legs before eating.

Range is now known to extend into southwest Canada.

Rafinesque's Big-eared Bat *(Plecotus rafinesquii)*

With their large, elaborate ears, members of the genus *Plecotus* are unique among eastern species. Ear size and a hoverlike flight identify members of three vespertilionid genera—*Plecotus, Euderma,* and *Antrozous*—as bats that can pick insects from the ground, rocks, or vegetation.

The outer ear and the tragus—its sentry-like appendage—are thought to serve bats as direction detectors. In *P. rafinesquii* the tragus stays erect when the bat is at rest, but the ears, particularly in winter, are coiled and often tucked beneath the wings. When awake, the bat greets a disturbance in its cave by turning its head about and waving its ears as though tracking echoes. It seems to peer at the intruder attentively.

Of medium size, gray to brown on its back, lighter on its belly, Rafinesque's big-eared bat weighs about a quarter to half an ounce (7 to 14 g). The female (above) is often a little heavier than the male. The range arcs across the southeastern United States from southern Virginia and central Indiana south to Florida and the Gulf states as far as eastern Texas. The bats form small maternity colonies that rarely exceed 200 members.

Originally, this species probably liked cave entrances year-round, roosting and hibernating in or near this twilight zone. Thus, it was among the first in jeopardy when people began disturbing caves. Once, many small colonies existed in ceiling domes or other heat-trapping spots of the twilight zone. Now displaced survivors most often roost in old buildings. In 1970 one of the last known cave colonies numbered 50 adult females. Since the cave owner has begun to protect it, the population has grown fourfold. Other small cave colonies have been found, but they are rare.

Pallid Bat *(Antrozous pallidus)*

Pale, sandy-colored fur cloaks this bat and merges it with its desert surroundings in western North America. In summer, it favors day roosts in rock crevices or buildings.

Emerging late in the evening, the pallid bat usually flies near the ground. It seems to hover, and it lands often. Uniquely, it feeds almost entirely off the desert floor, preying on crickets, grasshoppers, beetles, now and then a lizard or (above) a scorpion. During the night it hangs at rest in open, sheltered spots such as bridges, porches, or rock overhangs. Where it winters is uncertain—a few hibernation sites, similar to summer roosts, have been found.

This is a large bat—it weighs up to an ounce (28 g)—and it is a fairly noisy one. It produces a thin, high-pitched "squabble" note; it makes a clear, resonant call to other pallid bats; and, when frightened or angered, it bares its teeth and buzzes.

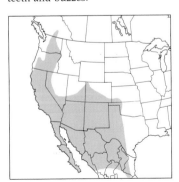

Family Molossidae

A tail that projects far beyond a short interfemoral membrane characterizes these bats and is the source of one common family name—free-tailed bats. Another—mastiff bats—comes from the heavily jowled, doglike face. Gray, brown, rusty red, or black, the insectivorous molossids number 16 genera and about 86 species. Their weight varies from 0.2 to 8.8 ounces (6 to 250 g). With their long, narrow wings, they are the fastest of all bats. Some molossids can fly 60 miles (96 km) an hour.

Members of this family generally inhabit the warmer regions of the world. They range throughout the Southern Hemisphere and also extend into southern Europe and parts of the western, central, and southern United States.

Molossids are the most gregarious of all bats and roost densely over large parts of cave walls and ceilings. In one instance, a researcher even reported a multilayered colony of these creatures.

Western Mastiff Bat (*Eumops perotis*)

Large ears joined across the skull and projecting over the eyes mark the western mastiff, the largest molossid in North America. *E. perotis* weighs about two ounces (57 g). Its size coupled with its narrow wings makes this bat, like others of its genus, a less than agile flier, though it can fly fast and cover great distances. Unable to take off from flat places, it must live in cliff crevices and other high roosts from which it can drop and thus launch itself. Airborne, the western mastiff bat climbs high in the sky. As it flies, it utters cries so loud they can be heard 1,000 feet (300 m) below. It feeds mostly over desert, at times over forest, on rather small insects, including bees and flying ants.

The mastiff bat's distribution is patchy. It occurs in central South America, in Cuba, and along the Mexico-United States border and north along the coast to San Francisco. This North American population roosts in buildings wherever a burgeoning human presence has preempted natural roosts.

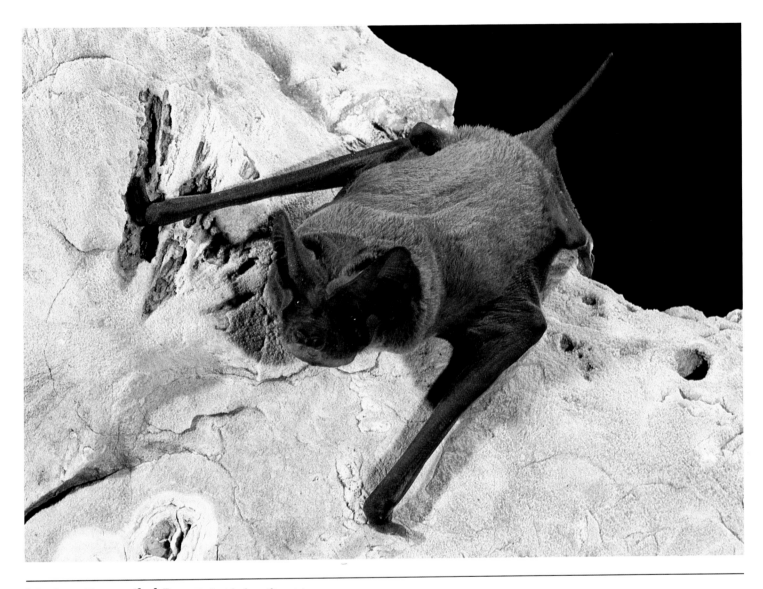

Mexican Free-tailed Bat (*Tadarida brasiliensis*)

The smallest of our molossids, *T. brasiliensis* weighs up to half an ounce (14 g) and is reddish brown to black. Just after sunset, millions of these bats come pouring out of their caves. They spiral upward like a column of smoke before dispersing to hunt insects on the wing. By dawn they return, dropping from the sky in free-fall to the cave entrance.

In June the female bears one baby. Its well-developed toenails adapt it to cling to the ceiling of the cave. But if a bat falls, beetles of the genus *Dermestes* may devour it. The eggs of this insect lie on the guano-covered cave floor through the winter. In spring when the migrating bats return, their fresh droppings stimulate the eggs to hatch. As larvae, the beetles feed on fresh guano and fallen bats during the summer. In the autumn, when the bats leave, the beetles lay eggs and then die.

The colony at Carlsbad Caverns

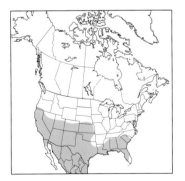

in New Mexico, once estimated at nearly 9 million, now numbers only 500,000. Insecticides have taken their toll. They reduce and contaminate the bats' food supply. As bats burn their fat stores, during arduous migrations as far as 1,000 miles (1,600 km), the toxins they have assimilated pass from fat to brain tissue, where they threaten the bat's life.

OVERLEAF: *In twilight flight, Mexican freetails swarm from their Texas cave, as the population of 20 million makes its nightly exodus.*

73

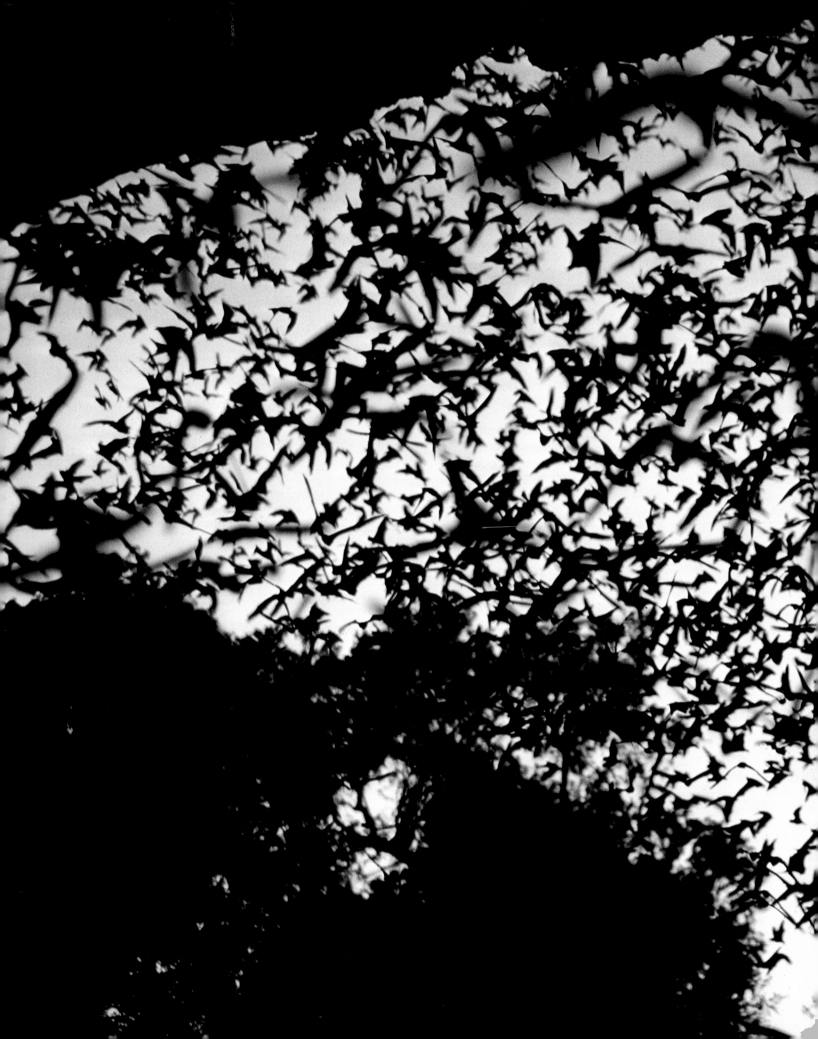

The Armadillo

ORDER Xenarthra

In profile, an armadillo shows one of its names: long-nosed. Of southern heritage, the animal adapts to cold by lining its burrow with leaves and grass—carried while hopping on its hind legs.

ARMADILLOS—"little armored ones"—are the only living mammals with bony shells. This jointed armor—actually modified skin—is overlaid with thin scales. In our North American species, the armor covers the top of the head and the back and sides of the body, and completely encircles the long tapering tail. Although the shell, or carapace, is generally comparable to that of a turtle, the armor is warm with body heat and flexible, moving with each breath. Sparse hair sprouts from between the shell's bony parts and on the animal's unarmored underside.

The armadillo's order was previously known as Edentata, meaning "toothless," but this characteristic strictly applies only to other xenarthrans: the anteaters of the New World. (The order's other survivors are the toothed tree sloths of Central and South America.)

The only xenarthran found in the United States is the common long-nosed armadillo of the family Dasypodidae, which has been given the scientific name *Dasypus novemcinctus*. *Dasypus* means "rough-footed"; *novemcinctus*, "nine-girded," refers to the movable bands about its middle.

One common name, "nine-banded," does describe armadillos found in the United States but not those of the same species that are found in the southern portion of its range. For in Argentina, Paraguay, and southern Brazil an armadillo usually has eight, not nine, movable bands.

Since nine-or-eight-*cinctus* is the most abundant and most widely distributed member of the genus, "common long-nosed armadillo" would seem to be a more appropriate name than "nine-banded."

Most who have sampled the armadillo's tender flesh say it equals or even excels chicken or pork in flavor. In parts of South America range-riding gauchos treat armadillos as traveling lunch boxes—to be thrown into a campfire and roasted for a meal.

This newcomer to the United States has spread from Mexico since the mid-19th century. People helped extend its range by transplanting it, possibly as potential food or as material for curios. Tourists are offered the result in the form of baskets or pocketbooks.

For whatever reason, the armadillo was deliberately introduced into Florida in the 1920s and, mostly on its own, proceeded to become established over much of the state. Around the same time, Louisianans began noticing large numbers of armadillos. They are now found throughout that state.

Climate determines the armadillo's wanderings. The northern border of its range varies according to changes in the severity of winters. Along their western border in Texas, armadillos have been retreating since 1860 because of a long-term dry-weather trend.

A creature of tropical to warm temperate climates, *D. novemcinctus* shivers in what would seem to us to be merely cool weather. Yet our armadillo is a rugged explorer compared with other members of its family and order. Only cold or very dry conditions have limited the range of this adaptable animal, which, as an omnivore, will eat almost anything anywhere it can find a home.

Nor has geography impeded the armadillo's progress. To ford a small stream, it may simply hold its breath and walk across the bottom, weighed down by its heavy shell. Faced with a wider stream, it inflates its stomach and intestines and floats across.

In South America, where the armadillos arose in the dawn of the age of mammals some 60 million years ago, 20 species survive. They range from a pink armadillo about seven inches (17 cm) long to a giant that weighs up to 120 pounds (54 kg).   RALPH M. WETZEL

Common Long-nosed Armadillo *(Dasypus novemcinctus)*

Most female armadillos have one or two young and two mammaries. But *D. novemcinctus* females have four teats—and replicate themselves in sets of four. The quadruplets, nearly identical and of one sex, are produced by division from a single fertilized egg. They later develop individual umbilical cords and placentas. More than

three months may pass before the embryos attach to the uterus and continue their growth. This delayed implantation may be an evolutionary adjustment of birth so that young can be born in the season that gives them the best chance for survival. Armadillos, as

they moved northward, probably fine-tuned the process so young were not born at the start of winter. Soon after their birth in February, the quadruplets venture out of the family burrow to forage with their mother (as the seven-month-olds do above).

Their foraging through leaves or grass litter is a snuffling, grunting

search, punctuated by frequent short, exploratory digs. A keen sense of smell helps them locate the insects, spiders, millipedes, pillbugs, earthworms, and snails that are the staples of the armadillo diet. Flicking out its sticky tongue, *D. novemcinctus* may snare 70 ants

Scant hair haloes the shell of an armadillo nosing for food. This swimming armadillo is inflated by swallowed air. An armadillo can hold its breath as long as six minutes to walk across a riverbed.

in a swipe—and gulp down 14,000 in a meal.

Because of its weak jaw and tiny, peg-shaped teeth, its food must be relatively soft or so small that it can be swallowed whole. The armadillo has been blamed for destroying garden harvests when it actually was consuming the leftover feasts of animals with more vigorous jaws, such as rabbits or raccoons.

When alarmed, an armadillo can run and dodge rapidly, making good use of its protective armor to charge through brambles and cactus clumps. It can also dig rapidly enough to keep ahead of a pursuer with a shovel. Cornered on the surface, it will leap straight up trying to escape. It apparently does this when straddled by a car, and that may be why many armadillos killed on roads have battered, not wheel-crushed, backs.

The small burrower—measuring, tail and all, no more than 39 inches (99 cm) and weighing up to 18 pounds (8 kg)—is unfairly regarded as a pest. But it does more good as an insect eater than damage as a garden digger.

Much more susceptible than humans to the bacterium that causes leprosy, armadillos are being used as a prime source for growing the bacterium in laboratories.

Rabbits, Hares & Pikas

ORDER Lagomorpha

A black-tailed jackrabbit, senses tuned to danger, eyes an intruder. Though fleet and alert, rabbits and hares usually die as prey. Only their remarkable fertility preserves the lagomorphs from extinction.

NO GROUP OF MAMMALS has established its place in American folklore as firmly as the animals that scientists call lagomorphs and other people usually call rabbits. As children, we first meet them in fairy tales, cartoons, and stories about Brer Rabbit, Bugs Bunny, and Peter Cottontail. Many superstitious people carry a rabbit's foot for luck, and children wait for the Easter Bunny—customs that date back hundreds of years.

The word "lagomorph" literally means "hare-shape." Though these small, furry animals were long considered rodents, the fact that they grow two pairs of upper incisor teeth, one pair directly behind the other, clearly sets them apart from rodents, which have only a single pair of upper incisors. The lagomorphs' secondary, smaller incisors, called "pegged teeth," snub against the backs of the large ones. But what purpose or function they serve remains unknown.

The fossil history of lagomorphs reveals that they probably descended from the condylarths, primitive hoofed mammals believed to be the ancestors of many herbivorous animals. Thus, in a sense, rabbits, hares, and pikas are more closely related to deer and elk than they are to rodents. The order Lagomorpha dates to Paleocene times, 65 million years ago, while lagomorphs achieved their present form during Oligocene times, 38 million years ago.

Lagomorphs are divided into two families, the Leporidae and Ochotonidae. The leporids of the Western Hemisphere consist of three genera: the hares of the genus *Lepus;* the rabbits or cottontails of the genus *Sylvilagus;* and the unique volcano rabbit *(Romerolagus diazi)*, an obscure animal found only on high volcanic slopes in central Mexico.

Hares are characteristically longer-legged and larger-eared than rabbits. Their keen hearing, enhanced by large, independently swivelling ears, enables them to detect predators at a considerable distance. Hares, which usually have much larger home ranges than cottontails, often venture far from cover, for they rely on their speed and agility to escape danger.

Hares also differ from rabbits in the way they bear and rear their young. Both hares and rabbits produce relatively large numbers of young. However, hares build almost no nest at all; they merely scrape out a depression in the dirt or sand. Cottontails often build elaborate nests carefully lined with grass and with fur the mother plucks from her undersides. The young of the hare are born fully furred, with eyes open, and able to move about, a condition called precocial. In contrast, the young of the cottontail, born naked and with eyes closed, must be cared for in a nest. These young are termed altricial. Hares may care for their young for only a few hours, a day—or not at all. Rabbits tend their young for as long as two weeks.

Hares and rabbits undergo complex reproductive cycles. The testes of the males, unlike those of most other mammals, are located forward of the penis and remain concealed within the body most of the year. Then, triggered by hormonal changes, they descend into the scrotal sac about a month before the onset of the spring breeding season. Thus the males are sexually primed several weeks before the females become receptive. The leaping, cavorting antics of these overzealous and unfulfilled males probably account for the expression "mad as a March hare."

During the breeding season, male hares and rabbits usually become very aggressive, battling one another to establish dominance. They may even kick and bite females.

Most of the breeding is done by a few of the dominant males.

Once the female becomes receptive, the act of copulation is required to induce ovulation of her eggs. In some species, as many as 30 percent of the fertilized eggs and fetal rabbits may die within the womb. These are taken up by the mother's body, a phenomenon called resorption.

Most native hare species molt and grow new white fur during the autumn; in areas of heavy snowfall this protective coloration helps them escape predators. Rabbits and pikas do not turn white in the winter.

The pikas, in the family Ochotonidae, are unusual among North American lagomorphs. They are found only in boreal habitats—cold, mountainous areas, such as the high Rockies and Cascades. Pikas reside almost exclusively on rocky slopes and, like all lagomorphs, are herbivorous. Pikas cut and store food for the winter and are active throughout the year. They differ most noticeably from the Leporidae in that their hind legs are similar in length to their forelegs and they do not have the hopping gait of the Leporidae. Pikas' ears are proportionately much smaller and their tails are not visible.

North American lagomorphs are ubiquitous. Whether it be the Arctic Circle, the Rocky Mountains, the deserts of the Southwest, or the marshes of Florida, lagomorphs of one species or another are ever present. The ability of these animals to adapt to and utilize diverse habitats has long intrigued mammalogists. For example, lagomorphs can obtain what little water they need from vegetation. Thus, from the coldest regions to the hottest deserts, lack of water does not hinder them. The only noticeable habitat requirement for most hares and rabbits appears to be abundant low vegetation for both food and cover. They may eat almost nothing but succulent green vegetation in the spring and summer and then in winter switch to twigs and blackberry canes. The change in diet does not seem to bother them.

Pikas, on the other hand, require a boulder-strewn habitat for shelter and for protection against predators. And the boulders must vary in size—with large and small chinks between them—so that the pikas can squeeze through openings too narrow to admit pursuers. The boulders must also be near a source of food, preferably an alpine meadow where grasses, shrubs, and other vegetation grow in abundance.

Hares and rabbits often "freeze" in a moment of imminent danger, as when a predator is about to strike. Why? Apparently

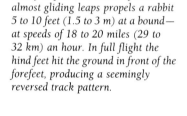

Bunny hop: A succession of low, almost gliding leaps propels a rabbit 5 to 10 feet (1.5 to 3 m) at a bound—at speeds of 18 to 20 miles (29 to 32 km) an hour. In full flight the hind feet hit the ground in front of the forefeet, producing a seemingly reversed track pattern.

in order to hide the visual target the predator may have been following: the cottontail's white tail or the jackrabbit's white sides. Suddenly deprived of the target, the predator may become disoriented and end the pursuit.

And, occasionally, a rabbit may do something unexpected—such as climb a tree. During one of my studies of brush rabbits in a pen, I noticed that an animal seemed to be missing from the enclosure. I searched several hours, concluding at last that it had indeed escaped. But as I left the pen, I happened to glance at a small fir tree next to the door and there, perched comfortably on one of the larger limbs, sat the missing animal.

Rabbits and hares may know how to evade predators but the skill ultimately does most of them little good. For their role is to await an inevitable fate at the bottom of the carnivores' food chain. Nearly all lagomorphs in a typical population die or are killed by predators in the first year of life.

Why then are hares and rabbits so abundant? The answer lies in their phenomenal fecundity. Females of some species may produce up to a dozen young per litter; some breed throughout the year. Cottontails may begin breeding when they are only three months old. An adult female can produce more than 30 young per year.

The ears of most lagomorphs are highly developed sense organs that may also serve as temperature-regulating mechanisms. The jackrabbit's large ears are laced with blood vessels that, when dilated, probably help dissipate the heat of the desert. The arctic hare, which certainly does not have to worry about being too hot, has ears that are much smaller than the jackrabbit's.

Both rabbits and hares have large, protruding eyes set high on their heads, giving them an almost complete circle of vision. But their eyes, like those of most nocturnal mammals, are probably blind to colors.

Rabbits, hares, and pikas reingest fecal material, an activity known as coprophagy. Some rodents and other animals also exhibit this behavior, which is not unlike the chewing of cud in such ungulates as deer or sheep. The excrement is reingested in the form of special glutinous pellets, which are usually produced in the evening or early morning. The pellets give the lagomorph certain vitamins it could not otherwise assimilate.

Most rabbits have a very small home range. They spend the greater part of their life on a few acres, or at most 20 to 25 acres (8 to 10 ha). The main determinants in the size of a rabbit's home range appear to be adequate cover, food, and nesting sites. Each species has

its own requirements. The marsh rabbit, for example, must live amid swampland reeds and rushes. The brush rabbit's primary requirement is low cover; the rabbit usually occupies a large bramble clump within its home range. Lagomorphs continuously use the same trails and resting places. This behavior makes their presence readily noticeable to those hunting or trapping them.

Many North American lagomorphs thrive in environments altered by humans—literally

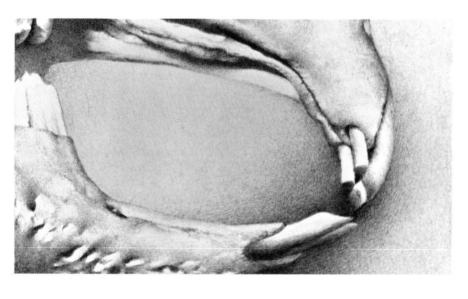

Dowel-shaped teeth buttress a lagomorph's chisel-sharp incisors—and distinguish members of this order from all other mammals. Rabbits, hares, and pikas chew by working their jaws side-to-side. The incisors, grooved in front, grow throughout the animal's life.

in our backyard. Because of their proximity to human beings, rabbits and hares are very important mammals. In the East, the cottontail is one of the most significant game mammals, providing millions of hours of recreational sport each year. However, in many areas of the West, cottontails and hares are considered crop depredators and spreaders of serious diseases. Many serve as hosts for the ticks that carry such diseases as Rocky Mountain spotted fever.

Widely scattered throughout North America, most cottontail species have stable populations and do not seem to be threatened by civilization. Only the pygmy rabbit and the New England cottontail appear to have populations and ranges so small that

mammalogists worry about the species. The range of the pygmy rabbit apparently is stable, but the range of the New England cottontail over the past several years has been noticeably diminishing. Originally distributed throughout the Appalachian Mountains, the New England cottontail currently appears to be restricted to small islands of suitable habitat. Wildlife managers are now beginning to consider measures for conservation of the New England cottontail.

Historically, the eastern cottontail frequented glades along river bottoms, while the New England cottontail, a much more secretive woodland species, ranged the higher elevations. But the situation began to change in the early 1900s, when the East experienced an apparent cottontail decline. To offset the decline, a massive restocking program took place throughout the Appalachians, with eastern cottontails being brought in from Kansas, Missouri, and other western and midwestern states.

The impact of the restocking program, which peaked in the 1940s and 1950s, has been tremendous. A native subspecies of the eastern cottontail, *Sylvilagus floridanus mallurus,* now no longer exists in pure form in the Appalachian region. Inbreeding with the western rabbits has completely altered the original native subspecies, genetically as well as morphologically.

These "new-breed" eastern cottontails, infused with fresh genetic vigor, have become highly efficient colonizers, spreading far beyond their old river-bottom haunts and adapting readily to changing ecosystems. Meanwhile the New England cottontail, unable to crossbreed between species, suffers because it cannot thrive in the face of steady encroachment upon its environment.

Will the newcomer eventually take over the shrinking range of the New England cottontail? Continued wildlife research will help answer this question—and the many others that are bound to arise when we and the lagomorphs find ourselves competing for space. JOSEPH A. CHAPMAN

Family Ochotonidae

Pikas do not have "hare-shape" bodies like other lagomorphs. Instead they resemble guinea pigs and have short, rounded ears. They lack a visible tail (although, like humans, they have a vestigial tailbone), and they scurry in a running gait similar to that of mice and other small mammals. Pikas are the only North American lagomorphs to cache food and, like hares and rabbits, they reingest their fecal pellets. Their young usually number two to four per litter and are born helpless and with eyes shut.

Unlike other lagomorphs, pikas are active mainly in the daytime. And they are quite vocal, chirping, bleating, and whistling as they scamper about their business.

Nor are they as aggressively territorial during the spring mating season as most other lagomorphs. Then, they frequently tolerate visits by other pikas—male and female. But they do become territorial when their pastures ripen. At harvesttime, trespassers are chased away—male or female.

Because most pika communities are isolated from one another, some 36 subspecies have developed on this continent.

Pika (*Ochotona princeps*)

Pika, piping hare, rock rabbit, mouse hare, cony, little chief hare. Call it what you will, this diminutive lagomorph thrives where most other mammals would starve or freeze to death. But the pika, admirably adapted to life at or above timberline, remains active even when snow lies deepest on its bouldery habitat.

A roundish body, short legs, and dense fur conserve body heat. The soles of the pika's feet are fur-clad too, for warmth—and for traction on slippery surfaces.

The pika gathers and stores food, scuttling over well-worn paths to harvest grasses and other plants that sustain it through the winter. It spreads the grasses in the sun to dry, slowly piling them into several bushel-size stacks. The owner jealously guards its caches, chasing off intruding pikas and marking each stack with urine and with scent from a facial gland.

A fully grown pika weighs 4 ounces (113 g) and reaches a length of about 8 inches (203 mm). Although 13 pika species exist worldwide, only *O. princeps* is found in North America. Most mammalogists now consider the collared pika of Alaska and northern Canada a distinct species, *O. collaris*.

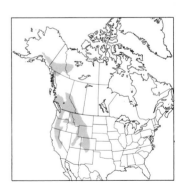

Family Leporidae

North American leporids are small-size to medium-size mammals of typical "hare shape." Except for two burrowing species (the volcano and pygmy rabbits), they make shallow depressions called forms in which they spend much of their time. In contrast, many Old World leporids dig elaborate tunnels and chambers known as warrens.

North American rabbits and hares generally are dark colored above and lighter below. Like the cottontail at right, they are extremely fleet-footed and, when running, often flash the showy undersides of their tails. They are nocturnal and crepuscular—active mainly at night and at twilight—and are considered important as game and as pests. Their breeding season lasts from a few months to year-round, depending on latitude and species. Hares and rabbits living in northern regions tend to have fewer litters— but more young per litter—than those living farther south. The females are polyestrous: sexually receptive at frequent and periodic intervals—including an hour or so after giving birth.

Leporid populations are subject to drastic fluctuations, typified by those of the snowshoe hare *(Lepus americanus),* which at times dies off in great numbers.

There are only three genera and about twenty species of leporids in North America.

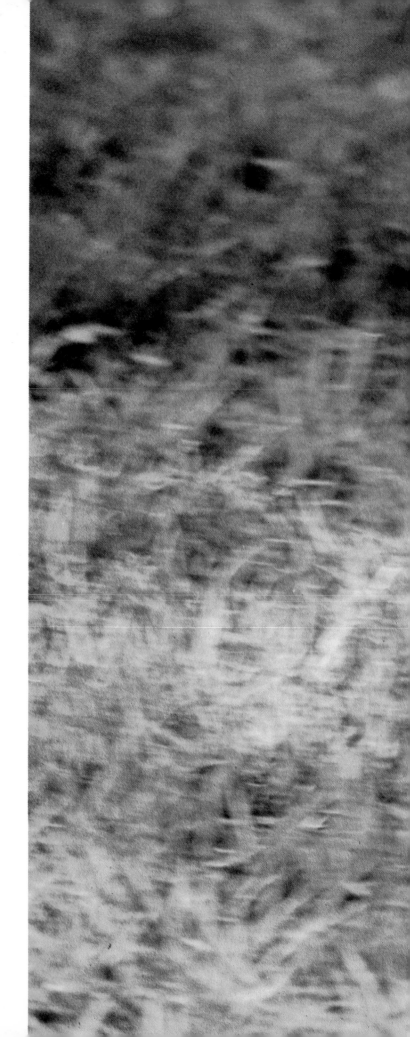

A black forehead blaze marks the secretive New England cottontail, distinguishing it from the more common eastern cottontail (left). Chromosome counts also differ between these two look-alike rabbits.

Eastern Cottontail *(Sylvilagus floridanus)*

Few creatures are better known—or more widespread throughout North America—than cottontails. All of them belong to the genus *Sylvilagus,* "woods hare," and most bear the distinctive emblem of their clan, the powder-puff tail. Among them, they have adapted successfully to habitats as varied as swamps and deserts, forestlands and prairies. And they are rabbits—smaller and stockier than hares, with ears and legs correspondingly shorter, and with their young born naked and helpless.

The eastern cottontail, widest ranging of the group, measures 15 to 18 inches (38 to 46 cm) long. Females (called does) usually are slightly larger and heavier than males (called bucks)—a trait common to all leporids.

A white or light-brown forehead blaze and a rust-colored patch at the nape of the neck mark this rabbit's brownish gray coat. (In contrast, the NEW ENGLAND COTTONTAIL, *S. transitionalis,* a separate species shown above at right, sports a black forehead blaze and black-rimmed ears.)

Eastern cottontails make their homes near brambly thickets or briar patches that offer protection against their many enemies. These shelters—called forms—are little more than scratched-out depressions in tall grass or weeds. Most American rabbits never dig tunnels or burrows. But they will borrow a burrow abandoned by

Eastern Cottontail

New England Cottontail (recently divided into two distinct species, *S. transitionalis* and *S. obscurus*)

another animal to escape bad weather or a foe in hot pursuit.

The average cottontail leads a very short life in the wild. Less than 25 percent of those born during any given season survive their first year. The rest are killed by parasites, diseases, and predators, including people. Hunters alone kill some 25 million cottontails a year.

A female eastern cottontail can raise up to six litters a year, each averaging five to seven young. Thus, if all her broods lived and reproduced, she would, at the end of five years, have established an empire of 2.5 billion bunnies.

Marsh Rabbit *(Sylvilagus palustris)*

Desert Cottontail *(Sylvilagus audubonii)*

The desert cottontail, also known as Audubon's cottontail, scratches a living amid the arid plains and valleys of the West. Slightly smaller than the eastern cottontail, it measures 14 to 16 inches (36 to 41 cm) from nose to tail. Its dark-tipped ears are longer and more pointed than those of its eastern relative.

Grayish brown fur and white undersides on tail and belly distinguish the desert cottontail. Its habitats vary from desert scrublands to grassy plains and stream bottomlands. Like all rabbits, it spends much of the day concealed in a shallow form and becomes most active after sunset. Grasses, shrubs, and other plant material, such as fallen fruits and nuts, make up the bulk of its diet.

When chased by a predator, the desert cottontail seeks the shelter

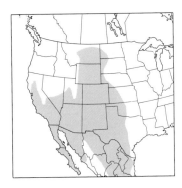

of rocks, shrubs, and burrows dug by other animals, including prairie dogs. And, like the brush rabbit of the Pacific coast, this cottontail occasionally climbs a tree—perhaps to better see its foes.

Any rabbit swims—if it has to. But some, including the marsh rabbit that inhabits the Atlantic and Gulf coastal states from Virginia to Alabama, take to water almost as readily as ducks. Naturalist John Bachman once kept a marsh rabbit as a pet. "In warm weather," he wrote, "it was fond of lying for hours in a trough of water, and seemed restless and uneasy when it was removed. . . ."

Clad in a coat of blackish to reddish brown, this 17-inch (43-cm) cottontail has a dingy gray tail. It conceals its nest amid the rank undergrowth of swamps, sloughs, and marshes, stepping along beaten paths on comparatively short legs. Reeds, tubers, and roots make up most of its diet.

When alarmed, the marsh rabbit leaps for the water, where it can outswim even a large dog. One rabbit was observed going strong 700 yards (640 m) from shore.

Marsh rabbits also elude their enemies by floating motionless among water plants, ears tucked down, with just noses and faces above the water. "On touching them with a stick," Bachman wrote, "they seemed unwilling to move until they perceived that they were observed, when they swam off with great celerity."

THE SWAMP RABBIT *(S. aquaticus)*, a relative, ranges from Alabama to Texas and along the Mississippi River as far as Illinois. At up to 21 inches (53 cm) long, it ranks as North America's largest rabbit.

Curiosity brings a young marsh rabbit to its tiptoes—a stance that permits a better view.

Nuttall's Cottontail *(Sylvilagus nuttallii)*

Sagebrush plains, forested streams, and alpine meadows of the West are home to one of the country's highest-living leporids—the Nuttall's, or mountain, cottontail. This smaller, grayish yellow version of the eastern cottontail weighs 1.5 to 3 pounds (0.7 to 1.4 kg) and lives at elevations ranging from 6,000 to 10,500 feet (1,829 to 3,200 m).

A shy, timid creature, it seldom ventures far from the protection of rock crevices or brushy thickets. If surprised in the open, it dashes for the nearest shelter, rather than dodging erratically as most other rabbits do.

Nuttall's cottontail, like most cottontails, bears a snowy tail and molts only once a year.

Like the rest of its kind, it stays much the same color in the summer or winter. But during the winter molt, the rabbit's feet sprout long, coarse hairs that enable it to negotiate drifts and feed on the tender tips of partly buried conifers.

Daylight most often finds this cottontail concealed amid boulders or in a brush pile. Like other rabbits, it is most active at dawn, dusk, and at night.

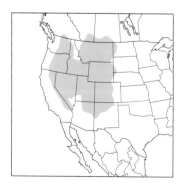

Pygmy Rabbit *(Sylvilagus idahoensis)*

North America's smallest rabbit weighs a pound (0.5 kg) or less and scuttles about the high, sagebrush plateaus west of the Rockies. Some zoologists question whether it should be considered a cottontail at all. To them, blood and other tests indicate that it should occupy its own genus— *Brachylagus,* "short hare."

Certainly this mite differs from other cottontail rabbits. For one thing, it has no cottontail. Its short, stubby tail is gray all over. Its ears are shorter and wider than those of most other cottontails and, unlike nearly all other North American rabbits, this one excavates its own burrows—or converts those made

by other small, digging animals.

The pygmy rabbit builds its den in relatively soft soil where sage and rabbitbrush form a dense, protective cover. It does not venture far from home and, if alarmed, scampers to one of several tunnel entrances concealed near the roots of a bush.

But the pygmy rabbit does share with cottontails a preference for nighttime activity. That's when it eats its favorite food—sagebrush.

Snowshoe Hare *(Lepus americanus)*

Large, broad hind feet matted with coarse hairs distinguish the snowshoe hare from the cottontail rabbit—and in winter enable the hare to bound nimbly across the deepest snows of its woodland domain. The hare's ability to change coats with the seasons—from summer browns and buffs to winter white (right)—gives it an alternate name, the varying hare. Crouched motionless in a snowdrift, the hare betrays its presence only with its dark eyes and black-tipped ears.

Weighing in at about 3 pounds (1.4 kg), the snowshoe hare, only slightly larger than a cottontail, ranks as the smallest of its genus. Its range extends across the brushy forestlands of Canada and into the eastern and western mountains of the United States.

Wild leaps and riotous chases mark its springtime courtship rituals. Females give birth to from one to nine young, known in their first year as leverets.

Snowshoe hare populations fluctuate widely over 10-year cycles. A variety of diseases induced by overcrowding are thought to decimate them periodically.

Black-tailed Jackrabbit (*Lepus californicus*)

Habitué of America's wide-open spaces, the black-tailed jackrabbit ranks as one of the most familiar small animals of the West. It makes itself at home in environments ranging from the desert scrublands of Mexico to the prairies and farmlands of western Missouri. Even the dunes and moors of Nantucket Island and Martha's Vineyard provide habitats for transplanted colonies.

Early settlers of the Southwest, noting the animal's extraordinarily long ears, dubbed it "jackass rabbit," a name subsequently shortened to jackrabbit. But this jack is no rabbit. It is a true hare. As such, it is lankier and leaner than a rabbit, has longer ears and legs, and its leverets are born fully furred and open-eyed. A large blacktail measures about 2 feet (61 cm) from nose to tail and weighs 5.5 pounds (2.5 kg).

As with all hares, blacktails rely on speed and camouflage for their defense. When flushed from cover, a blacktail can spring 20 feet (6 m) at a bound and can reach top speeds of 30 to 35 miles (48 to 56 km) an hour over a zigzag course.

When western settlers decimated such predators as the coyote and the kit fox, they also unwittingly triggered population explosions among jacks. Crops, orchards, and rangelands suffered. (Fifteen jacks can eat as much as one sheep.) Ranchers and farmers in some areas retaliated by herding thousands of hares to slaughter in wire-fence enclosures. One such drive in California netted 20,000 hares. Even so, the hardy blacktail still survives in droves.

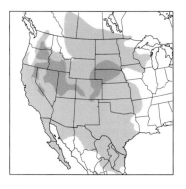

Black-tailed Jackrabbit
White-tailed Jackrabbit

This whitetail is not scratching; it's cleaning a foot. Hares and rabbits spend daylight hours preening, dozing—and scratching. In its white coat (below) it may burrow into a drift or lie in a scooped-out snowy form.

White-tailed Jackrabbit (Lepus townsendii)

Tails tell at a glance the chief difference between the white-tailed jackrabbit and its near relative, the blacktail. So do size and territory. Both species inhabit the plains and prairies of North America, but the whitetail ranges farther north, into Alberta and Saskatchewan. It is also slightly larger and faster than the blacktail.

There are other differences too, some not quite so apparent. The whitetail, unlike the blacktail, changes color with the seasons. As winter approaches the northern portions of its range, the long, buff-gray guard hairs of summer drop out and are replaced with white ones. Farther south, the whitetail's normal summer browns become grayish. But whether it lives near the northern or southern limits of its range, the whitetail keeps its black-tipped ears and showy tail unchanged throughout the year.

The whitetail is also a stronger swimmer than the blacktail. Both take readily to water if cornered, but the whitetail, kicking and paddling, makes, as one naturalist puts it, "rather fast progress."

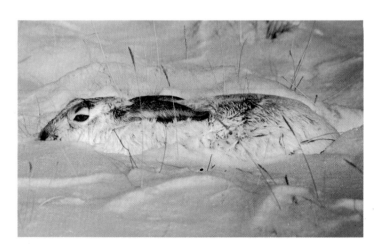

Antelope Jackrabbit (*Lepus alleni*)

The antelope jackrabbit ranks by all odds as the biggest, fastest, and flashiest of the western hares. Only one native American animal can outpace it in a dead run—its namesake, the pronghorn antelope (which, incidentally, is no more an antelope than this jack is a rabbit).

These big, rangy hares weigh up to 8 pounds (3.6 kg) and are covered with drab, brownish fur, except for a white vest along their flanks and undersides. Their ears, two enormous cartilaginous scoops, reach a length of 6 or 8 inches (152 or 203 mm) and make up a quarter of total body surface area. (By contrast, the ears of a blacktail comprise less than one-fifth of its body area.)

Antelope hares have been clocked at speeds of more than 40 miles (64 km) an hour, their powerful hind legs thrusting them 10 to 15 feet (3 to 4.5 m) at a clip. When startled, these hares bolt away in great bounding leaps, then settle into a rocking gait in which their hind feet hit the ground several feet ahead of where their forefeet had left the ground. At high speed they seem to skim effortlessly among the shrubs and bushes of their desert habitat, tacking this way and that just beyond reach of a pursuer.

With every jump, the hare displays a tantalizing target—a flashing patch of white hair on its rump. The hare apparently controls these displays: Only the side visible to the pursuer flickers on and off. These specialized erectile hairs, controlled by muscles in the back, normally lie hidden beneath the drab guard hairs. No one knows what purpose the flashing rump patch serves, but perhaps, as naturalist Ernest Thompson Seton once suggested facetiously, it warns the predator that the chase is hopeless. Dogs that have encountered an antelope jack have been known not to chase another hare of any kind for months.

The antelope hare, unlike the blacktail, does not take kindly to civilization. In the United States, it is now confined to southern Arizona's remotest deserts.

Arctic Hare *(Lepus arcticus)*

North America's largest hare tips the scales at 12 pounds (5.4 kg) or more, measures up to 27 inches (69 cm) from nose to tail—and is not averse to eating meat. Attracted by the bait in fox traps, these hares often trap themselves.

The arctic hare makes its home in treeless tundra. (A group huddles above for a sunbath.) A close relative, the Alaskan hare, *L. othus,* inhabits coastal Alaska.

Large, densely furred feet, short ears, and a chunky body encased in woolly white fur (which molts to brownish gray in summer at the southern reaches of its range) equip the hare for life in the frozen wastes. In winter, only its eyes and black ear tips show against the snow. Stout foreclaws dig through crusted snow for vegetation, and protruding, tweezerlike teeth can scrape moss and lichens from rocky crannies.

Mating takes place in early spring, often after sparring between bucks. Litters of up to eight are born in June or July.

OVERLEAF: *Arctic hares in summer dress gather conspicuously on Axel Heiberg Island. Groups of more than a hundred may gather to feed or rest in the Canadian high Arctic.*

Gnawing Mammals

ORDER Rodentia

WE RARELY SEE most of them, but they are all around us, day and night: the animals that belong to the order Rodentia. We know them as rats and mice, squirrels and chipmunks, woodchucks and prairie dogs. In numbers of individuals and species, this order is the largest group of mammals in North America—and on earth.

How many are there? We do not know. We can only guess that of the approximately 4,600 known species of living mammals, nearly 45 percent belong to the order Rodentia. Even though we may know of some of them by their common names, they are all rodents (from the Latin word *rodere*—to gnaw). For gnaw is what they all do to stay alive.

They live in more land areas than any other order—in forest and in desert, on farm and in city. Some places have more rodents than other places. California, Oregon, and Washington, for instance, have about twice as many species as are found in all of western Europe. And, though we think of them as small, they do come in many sizes. The South American capybara—a giant relative of the guinea pig—weighs about 100 pounds (45 kg). Some mice weigh less than an ounce (28 g). In North America, the largest rodent is the beaver. The smallest is the pygmy mouse, whose body is only two inches (5 cm) long.

Rodents have a single pair of upper and lower incisors—teeth adapted to cutting. Incisors can grasp, hold, and pierce objects, but most commonly the teeth are used to gnaw food. Each incisor is ever growing: Material is continually added at the base of the tooth, causing the tooth to be pushed farther out of the jaw. Each pair of upper and lower incisors forms part of a circle. As the teeth grow, they follow this arc. Thus, the upper incisors will grow downward, backward, and then upward into the jaw unless they are

constantly worn down. Rodents must keep gnawing to wear off the tips of their teeth. Antlers dropped by deer are often gnawed by rodents to obtain some mineral or simply to wear down incisors.

The enamel on the outer surface of the incisors is harder than the dentine of the remainder of the tooth. As the tooth wears, the tip becomes chisel-edged. This sharp edge makes the gnawing of hard objects possible.

In the mouth of a typical rodent, the grinding teeth are arranged into four rows, two in the upper jaw and two in the lower, just as in humans. In some rodents, the grinding, or cheek, teeth are, like the incisors, ever growing. In such species, roots never form on the cheek teeth. These rodents can eat abrasive material, such as grass. If a rodent with rooted—rather than ever growing—teeth ate that food, its teeth soon would wear away and it would starve to death. Thus, dental adaptation by some species, such as voles, has allowed them to eat grass and other food that many rodents cannot eat.

Most rodents—like humans—are plantigrades, which means they walk on the soles of their feet, with their heels touching the ground. Most species have five toes on their front feet. As in primitive mammals, such as shrews, the two bones of the lower front leg are not fused. Another little-known anatomical feature is the baculum, a structure of bone and cartilage that reinforces the penis. The males of most species of rodents—and many other mammals—have a baculum.

Many rodents, especially squirrels, hibernate. For some, the dormant state is short and shallow; for others, it is long and deep. The woodchuck, for example, may hibernate for nine months.

Some rodents, such as prairie dogs, are highly social and live by a code that may be complex. Others are territorial, living under a system in which individuals defend areas against members of their own species. Territorial kangaroo rats, if placed together, will fight to the death.

Rodents inhabit all parts of North America, from the upper limits of vegetation on the tallest mountains to the parched flats of Death Valley. The deer mouse probably has the most widespread distribution of any of our rodents: from the edge of the tundra southward

through many generations as species—and sometimes even genera—adjusted to changes in climate and vegetation. Today the hispid cotton rat, a tropical species, is expanding its range northward, probably in response to milder temperatures since the last Ice Age.

People have quite often surpassed nature in changing the ranges of many species. Norway rats and house mice, for example, were distributed throughout the world, primarily as unwelcome passengers on sailing ships. Fox

Portraits in sinew and bone show how anatomy divides the order Rodentia. Skulls' architecture and the rigging of masseter, or chewing, muscles sort rodents into three subgroups: the sciuromorphs, the myomorphs, and the hystricomorphs. The sciuromorphic ("squirrel-like") group is made up of two types. One has a muscle-jaw design found in primitive rodents, the only living example being the mountain beaver. The other type has the advanced arrangement that squirrels have. The jaw muscles attach to what would be cheekbones in humans. In the myomorphic ("mouse-like") rodents, a strand of muscle passes through the eye socket, forming a distinctive profile. The hystricomorphic ("porcupine-like") rodents have large skull openings for a massive jaw muscle. Differences in skulls arose from evolutionary changes in jaw muscles. The way a rodent's jaws work determines its habitat—and thus its diet. Because the lower jaw is loosely attached, rodents can chew in a rotary motion. This enables them to break down fibrous food.

Sciuromorph (mountain beaver)

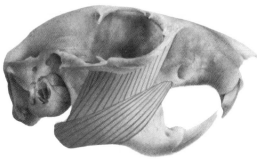

Sciuromorph (squirrel)

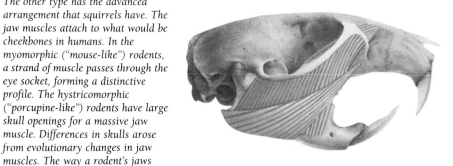

Myomorph (woodrat)

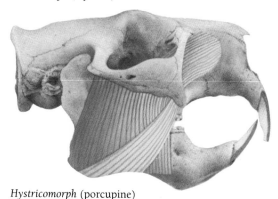

Hystricomorph (porcupine)

through most of the United States and well into northern Mexico. Within this area, deer mice may be found in every kind of forest, in grasslands, and in deserts.

By comparison, the Texas kangaroo rat lives only in mesquite grassland covering no more than 6,000 square miles (15,540 km) in Texas and Oklahoma, and the Cumberland Island pocket gopher is restricted to the sandy soils on its island off Georgia. Such distributions have been developed by rodents

squirrels and gray squirrels of the East have been introduced into the parks of many midwestern cities, simply because people like to see them there.

Thirteen-lined ground squirrels, historically of the Midwest grasslands, have enlarged their range by finding new habitats in the mowed strips along modern highways. These stretches of 20th-century grasslands have become avenues into the forested East for species that once lived only on prairies.

Prairie dogs have not been as lucky. Because their towns were on land that ranchers wanted, the prairie dogs were nearly wiped out, usually by vast poisoning campaigns. The habitats of many other rodents were destroyed as loggers felled forests and farmers plowed under grasslands.

Rodents move around in a variety of ways. Most of these types of locomotion will be familiar, but their technical names may not be. Each term conveys a great deal of information, which not only tells how an animal moves but also indicates something about the construction of its body.

Most rodents have relatively generalized skeletons that make the animals well adapted for walking, or *ambulatory* locomotion. These rodents can run (*cursorial* locomotion), but they are not highly modified for this type of movement, as deer and pronghorns are. Some rodents, such as kangaroo rats and jumping mice, have enlarged hind feet, elongated hind limbs, and small front legs. These animals are *saltatorial*—they progress by a series of leaps, with the main force supplied by their powerful hind legs.

Rodents that burrow can be divided into two groups, based on their lifestyles. *Semifossorial* animals—such as ground squirrels, woodchucks, and prairie dogs—dig burrows or holes in which they live. But most of their activities are on the surface. Truly *fossorial* rodents stay within their own burrow system and rarely come to the surface. Fossorial pocket gophers show such modifications as greatly enlarged forelegs and feet—and stout incisors—for digging, and small ears and smooth, almost slick fur to ease passage through the narrow burrow.

Tree squirrels are *scansorial;* that is, they can climb and run in trees. *Semiaquatic* rodents include beavers and muskrats. For their life in the water they possess such adaptations as webbed feet, flattened tails, reduced ears, and water-repellent fur. The flying squirrels are termed *volant* (from Latin for "flying"), though they are not capable of true flight. Between the squirrel's front and back legs are membranes that, when stretched, allow it to glide from high places.

Rodents are of considerable economic importance to every one of us—particularly as pests that feed on our crops, stored grains, and trees. Some rodents serve as hosts for fleas, lice, and other parasites that transmit diseases to people or domestic animals. The classic example is bubonic plague, or "black death," which is transmitted by a flea that lives on the black rat. A similar plague is harbored in the

A round sleeper, this eastern chipmunk curls up in hibernation, slowing down its metabolism to combat winter cold and famine. Occasionally it will awaken to nibble on food it has taken to bed. Other rodents go into deep, unbroken dormancy that may last for months. In summer, a similar torpor, this to beat the heat, is called estivation.

western United States by ground squirrels, prairie dogs, and some voles. The disease, which is often fatal, has been transmitted to humans, probably by fleas.

On the positive side, the native beaver and muskrat, along with the introduced nutria, annually yield furs worth millions of dollars. Laboratory rats and mice have been used extensively in medical research on human diseases and in studies of basic biology. Sportsmen derive enjoyment from hunting squirrels, which, early in our history, were a valuable food resource.

Rodents have been called the "primary converters" in a food chain that begins with their converting grass into food and ends with their being converted into protein by hungry predators. Game managers usually look fondly upon rodents as "buffer species"—animals that deflect predators from more valuable species, such as deer.

Two questions are often asked of professional scientists studying the relationships of animals (systematics). Why do you kill and preserve more than one of a species for study? How and why do you distinguish between closely related species? To answer the first question, we must ask

look very much the same. The genus *Peromyscus,* for example, has about 60 species, some so similar that even scientists have trouble telling the species apart. Or the animal could be from a group that has never been given much study. (This would be true of many tropical rodents in developing countries where scientists are few.) Or it may be from a place where the fauna is not well known. For rodents, this would include many areas in the United States.

Tools of the gnawing trade are chisel-sharp incisors and, for grinding, cheek teeth set back in the jaw. Even the space between—the diastema—helps. A working rodent can tuck its lips into the spaces to keep from swallowing unwanted chips.

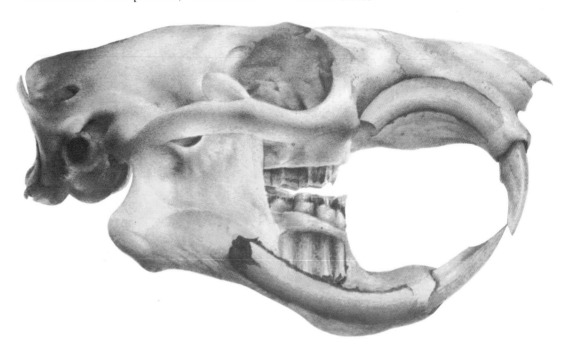

another: What can we learn from a single specimen? We should be able to determine what species it is, when the specimen was obtained, and where. We may be able to learn something about the biology of that individual specimen: Was it reproducing? What was its relative biological age? Was it changing (molting) its pelage (covering of fur)?

The determination of species may be difficult for several reasons. The animal may be young; it may exhibit some unique features not previously recorded. Or the animal may be from a group containing many species that

A basic principle of modern evolutionary biology holds that animals within a population are variable. Individual rodents of a particular species differ from each other much as members of human populations do. Darwin rigorously documented the variation within animal populations. It is this variation on which natural selection operates as it modifies and changes populations. This is the basic evolutionary process.

To document and understand variation in natural populations, a scientist must capture and kill specimens in a series. With a series of specimens from a single locality, we may obtain not only identifying information for a single specimen but also deeper understanding of variations within a population.

Like humans, rodents get larger as they mature. When a rodent's growth stops or is considerably slowed, the animal is scientifically termed an adult. It is with these adults that all other types of comparisons are made, because in adults there are no masking effects of growth. Many rodents, as they mature, will undergo two changes of their fur. Young rodents' fur may be quite different in color and texture from that of adults.

Males and females may be of quite different sizes. In species in which males vary significantly from females, all comparisons with other populations or species must be made sex by sex. Even among adults of the same sex within a population there will be differences between individuals. Some of them may simply grow larger than others or some may be lighter or darker than others. Populations of species of rodents may vary in appearance throughout their geographic range. To assess and understand this geographic variation, we need a selection of animals from all parts of their range.

A series of animals taken from selected sites throughout a range can thus tell us what species is present, exactly when and where it is present, and information about variations in age, along with sexual and geographic variations. Such knowledge is crucial in determining the true identity of a questionable species. And, once we know that identity, we are well on our way toward learning about the timing of reproduction, litter size, growth rates, the occurrence and pattern of molt, habitat selection, and geographic distribution of the species.

But why make such exhaustive (and exhausting!) studies? Some work that my colleagues and I did for the National Park Service may provide an answer. We were asked to survey the mammals of the Guadalupe Mountains National Park at the border of Texas and New Mexico. Within the park's

86,416 acres (35,000 ha) can be found mountain and desert, Douglas fir and yucca plant. In the oasis of an isolated canyon grow chinquapin oaks, hop hornbeams, bigtooth maples—trees that survived the region's ancient change from forest to desert.

Our assignment was to learn what mammals lived in this vast park, an ecological crossroads for fauna as well as flora. We were to determine which species were present and which were abundant, common, rare, or

A hapless woodchuck gives pathetic proof that a rodent's incisors never stop growing. Doomed by a malocclusion, it would die of starvation or a piercing of its skull. The photographer who found the woodchuck saved it from a lingering death.

endangered. We were also to make recommendations for their management.

An enormous group, genus *Peromyscus* with its 60-odd species—collectively called deer mice—was well represented in the park. Individuals were taken almost wherever we set out traps. We could have merely reported that the genus was widespread and should present no management problems. But we had to examine the finer fabric: How are the species of this genus distributed?

The first and probably most important job of a scientist who studies evolutionary relationships of animals is to be thoroughly familiar with all types of variation. In this case, we had to know the variations present in several series of *Peromyscus* that could be found in the area, including species in Texas, New Mexico, Arizona, and northern Mexico.

In studies of species relationships, we usually start with external characteristics. In *Peromyscus,* the size of the ears and hind feet is important for distinguishing many species. The tail is measured and its length compared with the length of the head and body. Other questions for identification include: Is the tail well haired? Is there a tuft of hair at the end? Is the tail one color or is it dark above and white below? What are the color patterns of the back, sides, and belly?

Next to be studied is the size and the shape of the skull. A number of standard measurements are taken on each skull, and these are analyzed with the aid of computers. The purpose of this exercise is to try to determine differences between species.

Some characteristics that proved particularly useful in studying the park's *Peromyscus* were the comparative lengths and shapes of certain tiny bones—such as those that form a mouse's nose. Another helpful clue was the male's baculum. This structure, which varies in size and shape from one species to another, is especially useful in distinguishing between closely related species.

Although all members of *Peromyscus* have the same number of molar teeth, the configuration of their cusps and ridges does differ between some species. These differences are difficult to detect and can only be understood after study of numerous specimens under a microscope. So we turned from computer to microscope. And we eventually had to increase magnification considerably, for we had to continue our search into the chromosomes of the specimens.

All members of the genus have 48 chromosomes, but they differ in the number of "bi-armed" chromosomes that they possess. Such chromosomes appear as an "X" under a microscope, and single-armed chromosomes appear as a "V." These chromosome preparations, made from bone-marrow cells, are studied under a microscope that magnifies nearly 1,500 times. Here, amid the carriers of the genes, we pondered another piece of the puzzle called species identification.

We identified seven species of *Peromyscus* in the park, including one that was a surprise. Each species had its own way of making a living and had found its own place to live. And so each species presented a different challenge to the park management. One species, the piñon mouse, had not been previously recorded anywhere in Texas. But the mouse had somehow discovered the stands of trees that would provide it with its favorite food: piñon nuts and juniper seeds.

The brush mouse, which likes dry and rocky habitats, lived in the high elevations of the park, where vegetation and fauna resemble those in the Rocky Mountains to the north. A very similar animal, the rock mouse, shared the habitat with its kin. The rock mouse has the Latin name *Peromyscus difficilis,* and descriptions of it often remark: "difficult to distinguish" from the brush or piñon mice which it so closely resembles.

High areas of the park are surrounded by and are in delicate ecological balance with the Chihuahuan Desert. In the more arid parts of this desert area we found that the deer mouse, namesake of the genus, was most abundant. Three other species completed our *Peromyscus* census. The cactus mouse inhabited rocky places along lower slopes. And the white-ankled mouse, as well as the white-*footed,* roamed scrub and grasslands.

Told exactly where these tiny creatures lived, park planners could put hiking trails or campsites where they would not disturb the habitats of species that needed protection. The piñon mouse would not be disturbed in its once secret hideaway. No animals would be endangered or accidentally wiped out.

The park still fulfills its mission of welcoming human visitors. Now another mission will be fulfilled: preservation. In Guadalupe Mountains National Park and in every other place where small creatures live, we are learning to preserve a vital national heritage by learning to share our habitats with the other mammals.　　HUGH H. GENOWAYS

Family Aplodontidae

This family consists of a single species, the mountain beaver (*Aplodontia rufa*), which, of all the rodent tribe, has a muscle and jaw structure like that of the most primitive rodents. It is the last of a group that traces its lineage back 50 million years to Eocene times. Its skull is a flat triangle; its cheek teeth grow throughout life.

The mountain beaver seldom ventures far from cover. Built for digging, it has a heavy, compact body, sharp claws, and small, gopherlike eyes and ears. Coarse whiskers guide the animal as it runs through dark underground tunnels.

The future of the mountain beaver is jeopardized by its slow birthrate, by shrinking habitats, and by its inability to adjust to human encroachment. Two subspecies along the California coast have been reduced to about a hundred individuals each.

Mountain Beaver (*Aplodontia rufa*)

The mountain beaver is not a beaver, nor is it confined to mountains. Under tangled thickets of bracken and thimbleberry, along streams, and in the meadows and forests of the Pacific coast, the chunky rodent tunnels for several hundred feet. When it surfaces, it leaves piles of dirt and rubbish near the entrance. Where trickling water makes the soil soggy, so much the better for digging. Burrow holes and tracks are all that most people ever see of this busy, mainly nocturnal animal.

The mountain beaver looks more like a stub-tailed muskrat. It measures 12 to 18 inches (30 to 46 cm) and weighs about 2.5 pounds (1.1 kg). Indians of the Pacific Northwest once wore robes made of its brown fur. The animal probably acquired a common name, sewellel, from the Indian word for the robes.

A strict vegetarian, the sewellel sometimes climbs trees to trim off needles and twigs. The rodent has been known to climb lodgepole pines to a height of 20 feet (6 m). Ascending—or descending head first—the sewellel steps on the stubs of cut branches as if on ladder rungs. Gardeners and tree farmers rate the rodent a nuisance. On one plantation it damaged 40 percent of the trees.

Females mature when they are almost two years old. The breeding period is only about six weeks long, and the number of young per litter is two or three.

A. rufa has a dubious distinction: It hosts a flea one-third of an inch (8 mm) long, the world's largest.

Family Sciuridae

"Shadow-tail," the meaning of Sciuridae, covers most of the squirrels. Exceptions are prairie dogs and some ground squirrels, whose stubby tails do not cast much of a shadow. Bushy-tailed tree squirrels are the chief exemplars of the family name.

Five dozen sciurid species are scattered from the Arctic to the desert Southwest. They nest in trees, dig burrows, tunnel into mountainsides, and sometimes invade attics. They tend to stay close to home, and some, such as flying squirrels and prairie dogs, live communally. (The four young arctic ground squirrels opposite practice territorial confrontation.)

All members have four toes on front feet, five on hind feet. Eyes set high on the head give ground dwellers like marmots a periscope view from their holes. They look out for predatory birds and meat-eating mammals. Human poisons and guns kill many. Automobiles destroy innumerable roadside foragers, such as woodchucks and chipmunks.

Storing as much as they eat, many squirrels hoard seeds, nuts, grasses, and fungi. Some also eat insects. Tree climbers sometimes raid bird nests.

Many species hibernate through cold months, lowering body temperature to near that of the burrow. Desert dwellers may aestivate, responding to intense heat by becoming torpid in their cooler dens. Dormant squirrels avoid not only cold or heat but also the dry conditions and scarcity of food that often accompany extremes in temperature.

Eastern Chipmunk (*Tamias striatus*)

Cheeks bulging with half a dozen acorns, the chipmunk darts down a hole by a fallen log. Front feet squeeze the jaws, and out pop the acorns, swelling a basement cache of nuts, seeds, and grain—a bushel (32 l) or more. (The *tamias* of its name means "a treasurer.") Preparing for winter, the chipmunk makes a nest of leaves right on the food hoard, then curls up for a long nap. But, unlike the fat marmots that hibernate continuously through the cold months, a trim, 4-ounce (113-g) chippie often wakes up hungry when blizzards blow—and snacks in its bed.

A shallow burrow, used for several years, grows in length—up to 30 feet (9 m)—and complexity, with branching tunnels, alternate exits, and extra chambers. Earth is hauled out in internal cheek pouches and spread far from the main entrance, which is camouflaged by leaves or rocks. Snug inside, the chipmunk plugs holes to seal off cold blasts and keep out predators that include dogs, weasels, snakes, and foxes. In the open, hawks and owls swoop down; house cats pounce. A loud "chip, chip"—even with cheek pouches full—signals alarm; a trilling "chip-r-r-r-r" says surprise. Sleek in racing stripes (hence *striatus*), it readily climbs a tree or swims across a pond.

Males court in early spring and again in early summer. A female may bite a suitor and drive him from her den. If they mate, four or five young are born 32 days later. In six weeks the young appear above ground and begin feeding on summer fare: berries, mushrooms, seeds, worms, insects, and gardeners' bulbs.

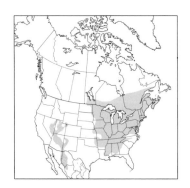

☐ Cliff Chipmunk
☐ Eastern Chipmunk

Least Chipmunk (*Eutamias minimus*)

It weighs but an ounce or two (28 to 57 g) and it measures but 7 to 9 inches (178 to 229 mm). Yet this chipmunk is least in size and name only. Most widespread of 21 western species, *E. minimus* thrives from alpine tundra to sagebrush flats. In the northern Great Lakes region its range overlaps that of the eastern chipmunk. Although the two have similar habits, the high-strung westerner is quicker on its feet. Another difference: Least and the other *Eutamias* species have two extra teeth.

When not eating a grasshopper or a caterpillar, a least chipmunk gathers and stores its food. In summer, where there are trees, it may lodge in an abandoned woodpecker hole or sleep in a high nest of leaves and grass. In treeless areas, it nests on the ground. High or low nests are vacated when frost drives least chipmunks underground. There they fall into a state of torpor. Sometimes in the dead of winter a least chipmunk

will wander about, leaving tiny tracks in the snow.

In spring, males seek mates, and five or six thimble-size young are born in May. For the next two months the young stay with their mother.

The YELLOW-PINE CHIPMUNK (*E. amoenus*) of the Pacific Northwest, a tree-climbing relative of the least, has a red coat and conspicuous stripes. The darker TOWNSEND'S CHIPMUNK (*E. townsendii*) ranges coastal forests from British Columbia to western Oregon. The COLORADO CHIPMUNK (*E. quadrivittatus*), one of several mountain species, lives in the Rockies.

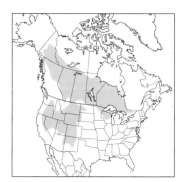

Cliff Chipmunk (*Eutamias dorsalis*)

A stranger to lush woodlands and gardens, this western chipmunk homesteads in dry desert country (map on page 108). Running ridge lines and scaling canyon walls, faintly striped, gray-backed *E. dorsalis* carries its bushy tail like a rudder. In a fall, the chipmunk swirls its tail and lands softly. A swaying tail signals alarm, in contrast to the flicking motions of ground squirrels.

Many enemies—four-footed, winged, crawling—hunt the cliff chipmunk as it forages for piñon nuts, juniper berries, insects, grass seeds, and birds' eggs. In dry spells, it eats juicy cactus fruits. Working

from dawn to dusk, it fills its internal cheek pouches with food, which it takes to a cache or to its nest in a cliffside niche or underground burrow. In raw weather, cliff chipmunks stay close to their nests of dried grass. They probably do not go into deep hibernation.

Early in the spring males seek mates. Four or five young are born after about a month's gestation. In six weeks or so, the family scatters, each member following the solitary ways of the species.

MERRIAM'S CHIPMUNK (*E. merriami*), a close relative, roams brushy and forested areas southward into Baja California. The GRAY-COLLARED CHIPMUNK (*E. cinereicollis*) favors pine-clad mountains in Arizona and New Mexico.

Woodchuck (*Marmota monax*)

Windmilling forepaws tear into the ground and hind feet kick out loose soil—a third of a ton (340 kg) or more—to dig a burrow. To Carolina Indians, this was *monax,* the digger. In light soil it can burrow out of sight in a minute. Its tunnel may meander 45 feet (13 m) or more at depths of three to six feet (1 to 2 m).

From the mounded entrance emerge eyes, ears, and nose—all placed strategically at the top of the head. Should a fox appear, it will shrill an alarm. Caught in the open, it dives for a hidden side entrance. It will scamper up a tree to escape

predators—or climb out on a limb for a ripe apple. (The fat chuck above picked one off the ground.)

When dense forest covered eastern North America, woodchucks lived along the edges of woods. As land was cleared, the edges increased—and so did woodchucks.

They raid fields of alfalfa and clover and forage along highways. Cars take a heavy toll, as do traps.

Solitary *M. monax* is usually aggressive and has no strong social ties. In late summer the four or five spring-born young are driven away by their parents, and they also soon separate.

Sticking close to home, a woodchuck grows lethargic and fat. In October it curls up in a grass-lined underground nest and hibernates until late February.

First to awaken is usually the male, a shadow of his peak weight—up to 13 pounds (6 kg). In March and April he sallies forth to nearby dens in search of a mate. Folklore borrowed from the Old World says that if he sees his shadow on February 2, he retreats to his den to wait out six more weeks of winter. Actually, to the groundhog, Groundhog Day is just another day for hibernating.

Hoary Marmot (*Marmota caligata*)

The hoary marmot ranges high country from Washington and Idaho to northern Alaska, its shrill notes ringing out from mountain meadows. French-Canadian trappers dubbed it *le siffleur*—the whistler. It also barks and hisses.

Favoring warm south slopes near timberline, whistlers fatten on sedges, lichens, roots, berries, and grasses. By digging burrows under boulders, they protect themselves from nearly all predators. But a hungry grizzly will tear up the earth to get a plump marmot.

Marmots may hibernate for as long as nine months. Mating is in May, and five weeks later litters

of two to five are born. Top weight for a mature hoary is about 20 pounds (9 kg).

The hoary is one of the most tolerant members of the genus *Marmota*. In its colony there seems to be no territoriality or strict dominance. Young may stay until maturity (two years) or longer. Unlike young in other species, they are not expelled by their parents.

The OLYMPIC MARMOT (*M. olympus*), brownish and similar in size to the hoary, is found in the Olympic Mountains of Washington.

Young hoary triplets appear to wear mittens. Caligata refers to Roman soldiers' black boots.

Yellow-bellied Marmot (*Marmota flaviventris*)

Rockchuck is another name for the yellow-bellied marmot. Both are apt. Home is usually a burrow under a rock slide, with a large boulder nearby to use as an observation platform. (From such a pedestal, the lookout above reveals its yellow undersides.)

These marmots live in colonies scattered at elevations up to 12,000 feet (3,650 m) through the Cascades, the Sierra Nevada, and the Rocky Mountains from British Columbia to New Mexico.

After a summer of gorging in alpine meadows, rockchucks retire to subterranean nests of dried grasses, hibernating from August to early spring.

Unlike the solitary woodchuck, yellowbellies usually form colonies

based on a dominant male with two or three females and young. (One male had a record harem of 31.) About a month after spring mating, a litter of three to six is born. After the birth of the next litter, the yearling young are driven out of the colony.

A full-grown yellowbelly weighs up to 12 pounds (5.4 kg) or more and is up to 28 inches (71 cm) long, with tail. Albinos are rare, but black-pigmented, or melanistic, marmots are seen frequently.

Arctic Ground Squirrel (Spermophilus parryii)

Arctic ground squirrels prosper in a frozen land. Ranging widely over the tundra, they pick their spots to fashion labyrinthine tunnels, usually three feet under the surface, above permafrost. As John

☐ Arctic Ground Squirrel
☐ Texas Antelope Squirrel

Richardson, a Scottish naturalist, noted, these squirrels enjoy living in society. (The muddy-faced digger above pauses, perhaps to look for neighborly help.)

Grasses, lichens, even caribou fur insulate hibernation chambers.

Curling into a ball, the ground squirrel spends about seven months underground. Although it rouses briefly at intervals, it does not fully awaken until April or May. Driven by the mating urge, it tunnels out through several feet of snow.

Five to ten young are born in late June. Growing rapidly, they near adult size of 2 pounds (0.9 kg) by September. They gorge on plants of the tundra—flowers, berries, grasses—scavenge carrion, and gnaw fallen caribou antlers. They beg around mining camps.

S. parryii—namesake of Arctic explorer Sir William Parry—is a staple of grizzlies, wolves, ermine, and other Arctic carnivores. Eskimos eat the squirrels and use the pelts to line parkas.

Texas Antelope Squirrel (Ammospermophilus interpres)

Scrub desert and rocky slopes west of the Pecos and south of the border are the domains of A. interpres—"the intermediary" between two close kin: Harris' antelope squirrel and the white-tailed antelope squirrel. All live up to their generic name: Ammospermophilus, which means "lover of sand and seeds."

When an antelope squirrel goes in search of what it loves to eat, it bounds across the dry sands, holding its tail protectively over its back. The tail has a white underside, and so, seen from the rear on the run, the squirrel looks a bit like an antelope. Close up, it's more like a chipmunk in size and appearance. A white stripe runs from rump to shoulder—not to the face, as on a chipmunk. Another

difference: The Texas squirrel's tail has a black band along each side.

On days when the temperature soars above 110°F (43°C), the squirrel spread-eagles on shaded ground to cool, or it retires to its den. It spends most days searching for mesquite seeds, cactus fruits, grasshoppers, and other fare. The squirrel in turn may become a meal for a rattlesnake or predatory bird.

The WHITE-TAILED ANTELOPE SQUIRREL (A. leucurus) ranges over the arid West from the central Rockies to the Mojave and Baja California. It does not hibernate, as many ground squirrels do, but in cold areas it grows a warm coat.

Thirteen-lined Ground Squirrel *(Spermophilus tridecemlineatus)*

With such a flag on its back, this squirrel would have captivated American colonists in 1776. But it ranges no farther east than Ohio, choosing to show its colors, shades of brown, in the midlands. It has more or less 13 stripes, the dark ones spangled with white spots.

S. tridecemlineatus—Latin for "thirteen-lined"—loves wide-open spaces. Farmland, roadsides, golf courses, and close-cropped pastures are prime sites for shallow burrows. Some meander 20 feet (6 m) or more. The squirrel scatters earth far from the main entrance and plugs side doors with vegetation.

S. tridecemlineatus is one of the few true hibernators. A pulse rate of more than 200 a minute slows to four or five. If body temperature falls below 32°F (0°C), the animal dies. But usually it awakens before freezing. It digs out in early spring.

When the long winter's sleep ends, males appear a week or two before the females. About a month after mating, females give birth to as many as 14 pink babies; seven to ten is average. Each weighs no more than a fifth of an ounce (5.7 g). Because the mother has only ten teats, some probably starve. Females have been known to devour young.

Seeds, roots, nuts, insects, worms, caterpillars, and carrion make up their diet. Some scavenge on highways and become carrion themselves. Other hazards are foxes, coyotes, skunks, snakes, and burrow-bulldozing badgers.

Uinta Ground Squirrel (*Spermophilus armatus*)

Ranchers in Idaho call this drab gray-and-brown ground squirrel "picket pin," a nickname it shares with several other species. When these squirrels sit stiffly erect on hind legs, they look like stakes driven into the plain.

But most of the time they are torpid underground. Members of a Utah colony spent an average of only 85 days a year above ground.

These strong swimmers live near water—not to swim but to eat succulent greens that grow there. (The diner above savors a dandelion in front of its burrow in Grand Teton National Park.)

RICHARDSON'S GROUND SQUIRREL (*S. richardsonii*) overlaps the range of *S. armatus* and extends from Colorado into Idaho, North Dakota, and Canada's prairies. The species are alike in appearance and habits. The main difference: The underside of Richardson's tail is buff; Uinta's is gray. In August both hibernate in burrows stocked with seeds, some dug from farmers' plantings. Uinta females bear four to six young. Richardson's has seven or eight, which are raised in females-only territories.

Golden-mantled Ground Squirrel (*Spermophilus lateralis*)

Golden locks that adorn head and shoulders give this squirrel its name. Campers in the Rockies call it the golden chipmunk. Though striped like a chipmunk, except for its face, *S. lateralis* is heavier.

It forages in alpine meadows and open stands of lodgepole pine, Douglas fir, and aspens. Seeds, berries, fruit, eggs, and insects make up most of its diet. And it accepts handouts from humans. It spends the summer collecting food and caching it in its burrow, perhaps under a rock or stump. (Bulging cheek pouches, as shown above, hold hundreds of seeds.)

In October, the squirrel curls up in a bed of dry grasses in its den and hibernates. It awakens in early spring, when males may scrap for mates. After gestation of a month, two to eight young are born.

California Ground Squirrel (Spermophilus beecheyi)

These burrowing squirrels at times connect their homes. One complex housed 11 squirrels, measured more than 700 feet (213 m), and had several dozen exits. Less elaborate tunnels serve squirrels that prefer to live alone.

The animals may be dormant in their burrows two-thirds of the year. They come out in early spring to renovate old dens with fresh grass, carving out new chambers to accommodate a growing family. (The mottled Californian above is on a food-seeking mission near the ocean.)

Adults—at 2 pounds (0.9 kg) they are heavyweights among ground squirrels—fatten on nuts, seeds, fungi, birds, and insects. Their taste for fruits and garden produce enrages growers, who fumigate dens and set traps. Health officials sometimes target these squirrels as bearers of bubonic plague and tularemia.

The ROCK SQUIRREL (S. variegatus), rivaling the Californian in size, ranges over the desert Southwest and deep into Mexico. It burrows under rocks and scrub, where it stores seeds and nuts. Unlike S. beecheyi, the dappled rock probably does not hibernate and is not colonial.

Black-tailed Prairie Dog (Cynomys ludovicianus)

Black-tailed prairie dogs work together to build a town, which they divide into neighborhoods. Then, from two to a couple of dozen neighbors get to know each other. If members of one neighborhood try to move into another, they are driven off.

Young and old "kiss" as a means of recognition (as above). But this family ritual ceases if a prairie dog senses danger. It barks a warning and dives into its burrow. At the same instant, hundreds of others heed the alarm and vanish underground. Predators stalk the dogs day and night. People lay traps and poison baits.

The foot- (30 cm-) tall rodent has been charged with crimes since Old West days: It ate grass needed to fatten steers; its burrow openings lamed horses. In fact, prairie dogs eat many weeds cattle won't touch—and burrow mounds rejuvenate the soil. Abandoned tunnels shelter owls and rabbits.

Millions of blacktails once lived on the Great Plains, excavating subways from Saskatchewan to the Rio Grande. Now prairie dogs thrive mainly on protected lands (range map on page 122). They spend the winter snug below ground. On warm days they venture out to forage for grasses and forbs. In spring, litters of three to five young are born—and are accepted into the neighborhood.

OVERLEAF: *Waist-deep in blooming wallflowers and tumblemustard, a mother blacktail and her curious young enjoy an outing in South Dakota's Wind Cave National Park.*

Weeee-oooo, whistles a blacktail, probably signaling all's clear. One of several distinct calls, it is given with the head thrown back and forelegs extended. During such displays, which may also express territorial rights or well-being, the animal often jumps straight up. Exuberant ones topple over backwards.

High-pitched, rapid barking means danger in a prairie dog town. A blacktail town crier (left) warns neighbors and kin to head for their dens. Underground, they will be safe from coyotes and bobcats, daytime hunters. But at night a hungry badger, following its keen nose, may rip out shallow tunnels to get at prey. A rare black-footed ferret will plunder a burrow.

Snakes sometimes hibernate in burrows and prey on pups. The prairie rattler (opposite) commands the attention of a blacktail half out of its hole. Should the snake move in, the prairie dog would not necessarily abandon its home. It might choose to wall off the occupied portion, tamping the earth plug with its blunt nose. Experts doubt that prairie dogs bury snakes alive, as folklore claims.

Vacant burrows attract such tenants as burrowing owls, mice, and salamanders. They, with the prairie dogs, are favored targets for eagles and hawks. A diving raptor, its talons extended, can terrify a colony. The ferruginous hawk above seized a prairie dog before it could find cover and now takes off with an interrupted meal.

A prairie dog's underground home provides shelter from storms, refuge from some predators, and a cozy place for a nursery. Mounds at entrances prevent flooding, serve as observation posts, and, being of different elevations, induce air flow through the burrow. Nesting and latrine chambers branch from the main tunnel. A notch, or wide place, near a burrow opening allows a prairie dog to turn around or to listen before venturing outside.

When two prairie dogs meet, they touch mouths as a sign of recognition. A blacktail pup (right) "kisses" its mother and paws her face, begging to be nuzzled. When prairie dogs aren't grooming each other, they are likely to be eating. A pear-shaped diner (opposite) minces blades of grass.

Douglas' Squirrel (*Tamiasciurus douglasii*)

Chickaree. Yellow-belly. Piney sprite. Sobriquets are many for Douglas' squirrel—namesake of 19th-century botanist David Douglas. Denizen of forests along the Pacific coast as well as rock slides above timberline, this was naturalist John Muir's "mockingbird of squirrels . . . barking like a dog, screaming like a hawk, chirping like a blackbird or a sparrow; while in bluff, audacious noisiness he is a very jay." (Given the chance, he will dine like a jay—as does the one above, pilfering peanuts spiked to a bird feeder.)

Food staples are cones clipped from pines, sequoias, spruces, firs. Some cones are husked on the spot for their turpentine-flavored seeds. But most cones go into storage, cached in moist places or buried in forest duff. The squirrels help reseed the forest—not only by what they inadvertently plant but also by collecting cones that foresters take for seeds.

The squirrels nest in tree holes, shredding bark to line nurseries.

Four or five young are born in May or June. Full grown in a year, they will measure up to 14 inches (36 cm) in length. Douglas' clan colors are olive (grayer in winter) above and bright rust below.

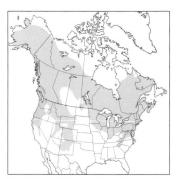

☐ Douglas' Squirrel
☐ Red Squirrel

White-tailed Prairie Dog (*Cynomys leucurus*)

Leaner, less social than the blacktails of the Great Plains, white-tailed prairie dogs favor mountain life and smaller colonies. These adults seem to squabble less than blacktail adults do.

Mining at altitudes up to 12,000 feet (3,650 m), they do not build mounds but usually pile the subsoil to one side. They do not need the extensive clearings vital to the more vulnerable flatlanders. Whitetails eat grasses and weedy plants, and, like the blacktails, prefer to have their food nearby.

The mountaineers spend a great deal of time eating; fat must be stored for a long sleep. By September most adults have retired to winter quarters underground. Hungry juveniles may tarry a few weeks, until heavy snows drive them to cover.

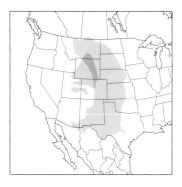

☐ Black-tailed Prairie Dog
☐ White-tailed Prairie Dog

Spring in mountain meadows heralds the breeding season. Pups are born in May. Few live to old age. Whitetails fall prey to hawks, eagles, badgers, bobcats, plagues—and man, the main enemy, who has poisoned, shot, gassed, and drowned them. Such campaigns have also hurt two other whitetail species: the Utah (*C. parvidens*) and Gunnison's (*C. gunnisoni*).

Red Squirrel (*Tamiasciurus hudsonicus*)

Falling spruce cones thump on the forest floor. But these are not windblown pods. Red squirrels harvest them green—scissoring them off above the stem—then store them in the ground, under rocks, or in holes to ripen. Below feeding-station branches, shucked scales pile 3 feet (1 m) deep and 20 to 30 feet (6 to 10 m) long. These middens often serve as cozy nest sites. But red squirrels will settle almost anywhere: a crow's nest, a woodpecker hole, a stone wall.

And they eat almost anything— nuts, berries, fungi, insects, larvae, shed antlers, eggs, young birds, sugar-maple sap. Syrup and pine pitch stick to forepaws, white vests, and rusty coats. Winter pelage has a drab cast.

Ojibwa Indians called the red squirrel *ajidamo,* roughly translated "tail-in-air" (for the reason shown above). Adults weigh about half a pound (227 g) and are a foot or so (30 cm) in length. Rarely silent, they chatter constantly and often scold intruders.

Male reds will chase each other during mating season, and fur may fly. Two to seven young are born in spring. There may be a summer litter. Overcrowding causes red squirrels to strike out across land and water. Tired swimmers have grabbed canoe paddles.

123

Fox Squirrel (*Sciurus niger*)

An early riser, the fox squirrel pokes its head out of its nest in a tree cavity. Scampering to the ground, it bounds across a woodlot for breakfast—a movable feast of hickory nuts unearthed and savored one by one. When corn is in the tassel, the squirrel severs an ear from a stalk and in solitary contentment gorges on the milky kernels. It will eat almost anything that grows: tree buds and roots, seeds, grain, fruits, berries, bulbs,

insects and their larvae, birds' eggs. Sometimes it taps a maple for the sweet sap and tender inner bark. (The thieving Everglades fox squirrel above raids a bird feeder in Florida's Corkscrew Swamp.)

During the day, the squirrel may stretch out on a limb to sun or nap. In hot weather its bushy tail curled overhead serves as a parasol. Rain drives it to one of several nests. High in a crotch of a tree it builds—in 12 hours if hurried—a nest of leaves and twigs. It favors a

tree-cavity nursery for its spring and fall litters of two to five young.

A full-grown fox squirrel weighs up to 3 pounds (1.4 kg) and is up to 29 inches long (74 cm); half of that is tail. It sometimes falls out of trees. But even after a drop of 40 feet (12 m), it has been seen to land on all fours and run back up.

It will run and hide from an intruder rather than stand and scold, as Douglas' and red kin do.

But it cannot escape hunters, who, prizing its meat, kill millions every year. Saws may harm them more than guns: Clearing hardwood forests robs the fox squirrel of food and shelter. Severely reduced in the Northeast, it has extended its range to Colorado and the Dakotas. Transplants thrive in West Coast parks and campuses.

When 18th-century botanist Carl von Linné—better known as Linnaeus—described the fox squirrel, he dubbed it niger, believing the black form (above) to represent the species. Though the name is kept for the species, we know now that this dark form, S. n. niger, is only one subspecies (a geographic variation within a species) of the fox squirrel. Such a distinction may occur because of isolation and adaptation to local environment.

S. n. limitis (left) lives mainly in Texas. Two races—the Everglades fox squirrel, S. n. avicennia (opposite), and the Delmarva fox squirrel, S. n. cinereus (upper)— have declined in numbers because logging has shrunk their habitats.

Gray Squirrel *(Sciurus carolinensis)*

Badge and banner of the gray squirrel is its bushy tail. In rain it curls over the back as an umbrella. In cold snaps it's a blanket. The squirrel flicks it in greeting, fluffs it in anger, waves it in courtship. It serves as a balance. In a fall it's used as a parachute; in water, as a rudder. Half of the gray's length of 15 to 20 inches (38 to 51 cm) is that versatile tail.

Common in parks and hardwood forests, wherever nut trees grow, *S. carolinensis* finds home in its namesake Carolinas and in the rest of the eastern United States, plus a slice of Canada. Its menu includes nuts, berries, seeds and flowers, insects, mushrooms (some poisonous to humans). But foremost are acorns, some of which are buried and forgotten. Many a stately oak began as part of an absent-minded gray's hoard.

Exploding populations tax food supplies and may trigger a lemminglike exodus. Many squirrels die on highways. Many drown. Others fall prey to owls, snakes, dogs—and the gun. Few live past five years, most live less than a year. But some in captivity have lived 15 years. So prolific are they—mothers wean two sets of triplets or more annually—that a population can tolerate heavy hunting. Leaf nests in trees, easily seen in winter, betray the squirrels' presence to hunters.

Pink eyes mark this gray as a true albino. Hundreds of such albinos roam protected in Olney, Illinois.

Western Gray Squirrel (*Sciurus griseus*)

Bounding through redwood forests and woodlands of oak and pine, gray squirrels of the West forage for acorns and pine nuts. These they store singly in shallow holes on the forest floor. Sensitive noses, not keen memories, relocate caches; overlooked seeds sprout.

Westerns readily climb and travel in the treetops. (Plumed tail steadies the one above, poised on a log.) Bulky nests of sticks and leaves lined with shredded bark house families of three to five. Enlarged woodpecker holes also serve as nurseries.

Young are born from late winter through spring after about six weeks' gestation. When fully furred, they have blue-gray coats and white undersides. Adults, measuring up to 24 inches (61 cm) to tip of tail, are 20 percent larger than eastern gray kin.

☐ Abert's Squirrel
☐ Gray Squirrel
☐ Western Gray Squirrel

Abert's Squirrel (*Sciurus aberti*)

Tassel-eared and bushy-tailed, Abert's squirrel cuts a dashing figure in sporty pelage. A most spectacular animal, it flaunts a silver-and-white plumed tail, white vest, and gray coat accented with russet and black. Tufts are distinctive earmarks in winter. Heavy for a tree squirrel, *S. aberti* weighs up to 2 pounds (0.9 kg). Ponderosa pine forests of the Southwest provide food and shelter. It husks pine cones for seeds, but its diet also includes the leaf buds of pines, carrion, bones, dropped antlers, weedy plants, berries, and fungi. (The forager above checks a fallen log for mushrooms.)

Nests of woven twigs sway in high branches, rocking spring broods of three or four. Melanistic individuals are common.

Kaibab squirrels, black-bellied with white feather-duster tails and long tasseled ears, live in a pocket forest on the north rim of the Grand Canyon. Isolated by desert and the mile- (1.6 km-) deep chasm, they are believed to be a distinct subspecies, not another species. Prisoners of geography, they face an uncertain future.

Antlerlike ear tufts mark this youngster as an Abert's squirrel. In summer, the handsome ears are not tipped by tassels.

Southern Flying Squirrel (*Glaucomys volans*)

At dusk, after the red and gray squirrels have retired, *Glaucomys volans*—literally, the flying gray mouse—stirs to life. Wide liquid eyes peer from an abandoned woodpecker hole to scan for cats or an awakening owl. When all's clear, the southern flyer emerges, its 2-ounce (57 g) body fitted into a cinnamon-tinged gray coat that seems several sizes too large. (Folds of loose skin, attached between front and hind legs, accordion on the squirrel above.)

From a perch high in an oak or other hardwood tree, the 10-inch (25-cm) squirrel cranes its head from side to side, as if calculating a glide path. Its muscles tense. With a vigorous thrust of its hind feet, it dives into space. The flanking membranes, taut between stretched limbs, create a planing surface of more than 50 square inches (323 sq cm). To change direction in mid-glide, it maneuvers arms and tail. Like a billowing sail, the slack membrane cups air. Braking for a landing, the squirrel lowers its featherlike tail, then jerks it up before touchdown.

Flying squirrels come out at dusk with the bats and, like them, sometimes hang by their toenails in tree cavities. For homesites, the flyers favor mature woods with woodpecker-tested trees. Also acceptable are leaf nests, attics, and perhaps a mailbox. Foods include acorns—stored and in season—larvae, seeds, berries, nuts, insects, fungi, and baby birds.

Spring mating produces three to five young after 40 days. Females in the Deep South may have a second litter in the fall.

The NORTHERN FLYING SQUIRREL (*G. sabrinus*) lives in evergreen forests in Canada, Alaska, and northeastern and western states. It may be twice as heavy and a few inches longer than *G. volans*.

Frisbee in fur, a southern flying squirrel sails in for a landing. Lateral membranes enable it to glide—not fly—up to 100 feet (30 m).

Family Geomyidae

Built for burrowing, geomyids—"earth mice"—are compact and have sleek fur, tiny ears and eyes, sharp incisors, and sturdy front legs with long, curved claws. External fur-lined cheek pouches—the gophers' "pockets"—carry chopped roots and tubers to underground storage chambers. (Below, empty pockets frame incisors, which are always exposed.) Vertical grooves—or their absence—on the yellow incisors help identify a genus.

North Americans dating from the Oligocene, family members are found from southern Canada to the edge of Colombia. Most wear shades of brown, but pelage ranges from whitish to almost black. They are active all year. Tunnels that serve as feeding troughs meander at grass-roots depths. Deeper excavations shelter nests. Except for two months with their mother as young, pocket gophers prefer to live alone.

Western Pocket Gopher (*Thomomys mazama*)

As a green plant sinks into the ground, a camper in Oregon rubs his eyes in disbelief. A trick of his imagination? No. He has witnessed an occasional feeding technique of pocket gophers. Burrowing just below the surface, they can pull entire plants underground. Sharp incisors cut the stems into pieces small enough to fit inside cheek pouches.

Cautious gophers seldom venture above ground to browse. When one does (*T. mazama* feeds above), it may stray no farther than a tail's length from the safety of its burrow opening. It may leave its burrow to mate, but it usually doesn't. A male will dig his way to a female, determine if she is receptive, and then mate underground. Afterward, the male goes his solitary way. The female, in a deep underground nest, gives birth about a month later to four or more babies. In some areas she bears a second litter the same year. Except for mating, these solitary rodents do not socialize. In fact, if two gophers of either sex happen to meet, they may fight each other to the death.

Pocket gophers live in the western half of the United States as well as the Southeast. They vary in length from 5 to 9 inches (127 to 229 mm), excluding tail, and weigh from 2.5 to 26 ounces (71 to 737 g). *T. mazama,* medium-size, has a range that includes Puget Sound, the Cascades, and northern California. Like all pocket gophers, it builds an elaborate burrow system, with a winding tunnel leading down to chambers for nesting, food storage, and waste. It digs with long front claws and upper teeth whose chisel edges cut through tree roots and even underground cables. Loose soil is pushed out with the head and forepaws into fan-shaped mounds.

Coyotes and badgers dig out gophers, but poisons and traps take heaviest tolls. Naturalist Jack Schaefer said the gopher "is the target of more means and devices aimed at his destruction than any other four-legged creature. . . ."

Family Heteromyidae

These are the "different mice"—about 60 species of pocket mice, kangaroo rats, kangaroo mice, and spiny mice. All have external fur-lined cheek pouches like those of pocket gophers, their closest relatives. Most of them have skulls bulging with oversize hearing chambers designed to pick up low-frequency sounds that predators make—the fanning of an owl's wing, for example.

Most species live in desert areas of the western United States and northern Mexico, digging shallow burrows in sandy soils. They forage at night and store seeds underground. Most do not drink water, but obtain moisture by metabolizing food. Sand or dust baths condition their silky coats. They are solitary except when mating.

Kangaroo mice and rats have tiny front legs and long hind limbs. They will jump to evade predators and each other; a pair will fight until one quits. Long tails serve as rudders for jumping, props for standing—and, for kangaroo mice, as an energy reserve. Unlike their jumping kin, pocket mice usually travel on all fours.

Although rugged and adaptable, heteromyids are rapidly expanding the ranks of endangered and threatened wildlife. Having survived severe climates and natural predators for centuries, many localized species and subspecies are being eliminated by urban development or conversion of their grassland habitats to agriculture.

Silky Pocket Mouse (Perognathus flavus)

Thumb-size with a two-inch (5-cm) tail, *P. flavus* scurries through the desert scrub for seeds. These it carries in cheek pouches to underground caches. It carefully shells seeds before eating them.

Pocket mice spend the day in their burrows, plugging entrances with dirt. They may become dormant in cold weather.

The warm winds of April usher in the breeding season, which lasts until November. A silky mother bears two or more litters a year of two to six young. But populations are not excessive; snakes and owls keep the mice in check. In Texas' Big Bend National Park, a study showed only 16 out of every 100 silkies lived for more than a year.

Some two dozen species of pocket mice occur through the West. The PLAINS POCKET MOUSE (*P. flavescens*) favors open country from the Dakotas to northern Mexico. Several species live in Death Valley—the LITTLE POCKET MOUSE (*P. longimembris*) at high elevations; the GREAT BASIN POCKET MOUSE (*P. parvus*) lower down; and the DESERT POCKET MOUSE (*P. penicillatus*) on the flat desert. Most authorities now place the last species and several related pocket mice in a separate genus, *Chaetodipus*.

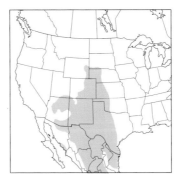

Merriam's Kangaroo Rat *(Dipodomys merriami)*

Not kangaroo, not rat, this denizen of the desert Southwest and Mexico is closer kin to the pocket gopher. The generic name implies it's a "two-footed mouse." It isn't. *D. merriami* and kind have 20 teeth (not 16 as in typical mice) and four functional limbs, the short front pair being useful for picking up grass seeds. These are stored in shallow burrows. One hoarder cached 14 bushels (490 l) of seeds. The starchy kernels provide water

The kangaroo rat can bound 18 inches (45 cm) high, swinging its tail to change directions. The 4-inch (102-mm) rodent has a tail 6 inches (152 mm) long.

It haunts the night but shuns a bright moon. It is solitary—and a chance meeting (like that of the two above) may lead to a spat. The solitary life is interrupted by mating once or twice a year. The one to five young may be born any time from May to October.

as well as food when metabolized, eliminating the need to drink.

Another key to survival is the rodent's keen hearing, four times sharper than a human's. A thin-walled skull and a middle ear larger than the brain enable the animal to "tune in" an owl's whirring wings or a sidewinder's scraping rattles. Such sounds trigger instant flight.

Skeletal illustrations compare the elongated hind foot of Merriam's kangaroo rat to a more typical rodent foot, a deer mouse's (lower). Partially fused bones add strength, enabling some kangaroo rats to leap as far as 10 feet (3 m).

Ord's Kangaroo Rat *(Dipodomys ordii)*

Found in 17 western states, two Canadian provinces, and Mexico, this species is the most widespread of kangaroo rats. Habitats include arid Kansas plains, Saskatchewan riverbanks, mesquite-covered sands of the Southwest, and sagebrush lands of southern California.

Unlike other species, *D. ordii* will occasionally drink water. But when sources dry up, the self-sufficient creature makes its own moisture by metabolizing seeds. It bathes in sand to keep body oils from matting its silky coat. Abroad only at night, the rodent eats insects and gathers seeds in its cheek pouches. It stores the seeds in chambers a few inches below ground. Plows and hoofs sometimes ruin the shallow burrows.

More likely sources of danger are rattlesnakes, owls, badgers, and coyotes. Like other nocturnal species, Ord's defends its home during the day by barring the doors—plugging entrances with sand. The drumming sounds from below, caused by its thumping hind feet, may scare off some intruders.

In the open, the kangaroo rat turns its back on danger—but with a vengeance. It kicks up a tiny storm, spraying sand in the predator's eyes. Soles of the long hind feet have a tread of hair, providing nonslip traction.

Solitary by nature, a kangaroo rat may greet one of its own kind with belligerence. It bounds into the air and flails its whiplike tail, which is longer by half than its body; together they may reach 12 inches (305 mm). A swift kick powered by a leap can send an opponent sprawling several feet. Only at breeding time—both spring and autumn in southern parts of the range—do kangaroo rats tolerate each other. The young—two to five in a litter—grow rapidly. Within two months females are mature enough to breed.

Banner-tailed Kangaroo Rat *(Dipodomys spectabilis)*

A heavyweight among kangaroo rats, this southwestern species weighs about a third of a pound (151 g) and measures up to 15 inches (381 mm) in length. More than half of that is tail, tipped with its distinctive white banner.

Nothing provokes a bannertail so much as another bannertail. They jump and twist (like the pair above), kick sand, rake each other with long hind claws, and bite when they get close enough. A kangaroo rat will fight rather than risk losing its hoard of seeds to a cheeky neighbor.

Working alone, a bannertail spends much of its nights enlarging and stocking its burrow. Several years of toil by successive generations results in a sandhill home up to 3 feet (1 m) high and 15 feet (4.5 m) wide. It may have a dozen entrances, the main one wide enough for a desert fox to get its nose in the door. But the tunnels narrow, and often dead-end. A trapped bannertail will burst through an escape hatch plugged with loose sand. In the open it can leap 10 feet (3 m).

A storage chamber may contain several pounds of seed. Bearded heads of grasses are nipped off at the stems and stuffed in cheek pouches, then carried underground to be stripped of the grain. Shredded grass lines nests where three litters of two young each may

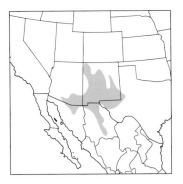

be whelped between January and August. In nine or ten weeks the young set out to find homesites. They favor gravelly mesas, sandy foothills, and shortgrass plains.

On hot days bannertails tamp sand in burrow openings so they can sleep in humid comfort. They emerge only at night, venturing out even in subfreezing weather. Rain, however, keeps them inside. So does moonlight—except during drought, when food is so scarce that even daylight won't deter them. True to their kind, these kangaroo rats avoid water, relying instead on their unique body chemistry to provide them with adequate moisture.

Dark Kangaroo Mouse (*Microdipodops megacephalus*)

On a warm cloudy night when the moon is down, this furtive creature is most likely to leave its home beneath the sand. If forced out of its dark den during the day, the animal instinctively darts for the shadows. Hopping on fringed hind feet like a tiny kangaroo, it bounds over the desert to forage amid sage and rabbitbrush for seeds of the blazing star, a desert flower. Dainty front paws gather the black seeds—plus insect morsels in summer—and pack them tightly in cheek pouches, furry saddlebags on each side of the mouth.

The seeds go into underground caches, for this rodent puts a little something away for a rainy day. It avoids wet weather as well as bright light—preferring a sand bath to a shower—and it seldom, if ever, takes a drink. Metabolic water satisfies thirst.

When drought limits seed crops, a kangaroo mouse hibernates—or aestivates—living off fat stored in its 3- to 4-inch (76- to 102-mm) tail. The animal sleeps on its back, with forelimbs stretched over the head and hind legs tucked against the belly. Half the 3-inch (76-mm) body is head (as witness the button-eyed male above), a fact that explains the specific name *megacephalus*—large head. *Microdipodops,* the generic name, roughly means "tiny two-foot."

The PALE KANGEROO MOUSE (*M. pallidus*), several shades lighter than its cousin, inhabits west-central Nevada desert and small adjacent areas of California. In each of the two species, spring litters average about four.

Mexican Spiny Pocket Mouse (*Liomys irroratus*)

"Plain mouse sprinkled with dew" is the meaning of this rodent's scientific name. Plain it is in drab tones, but the dew is poetic. More evocative is the "spiny" part of its common name, for its coat is sprinkled with stiff bristles. (Seen in the light, as above, they appear to sparkle . . . like the dew?)

L. irroratus shuns light, however, and stays all day in its burrow under a rock or bush. It comes out only at night to collect seeds and fibrous nest material. Although not a jumper like its kangaroo-type cousins, this rodent balances on hind feet to forage. It stuffs seeds into fur-lined cheek pockets that open outside the mouth and reach back almost to the shoulders. In the burrow, the pouches are emptied—turned inside out—and groomed.

At home in central Mexico and the southern tip of Texas, the 4- to 5-inch (102- to 127-mm) pocket mouse—with a tail of equal length—likes it high and dry. Desert, thickets, and wooded mountains are favored habitats. Unlike most members of the family Heteromyidae, this species probably requires an occasional sip of water, since laboratory studies indicate it cannot maintain body weight on seeds alone.

The females apparently breed throughout the year, with peak months of reproduction from August to November. Litters average four.

☐ Dark Kangaroo Mouse
☐ Mexican Spiny Pocket Mouse

Family Castoridae

The beaver, only member of this ancient family, outbulks every rodent but the capybara of South America. And the beaver outdoes every creature—except people—as a landscaper. It is so committed to life in the water that it spends half its time swimming and much of its energy damming a stream to create its own habitat: a beaver pond.

For these tasks the beaver is superbly adapted. Its paired incisors hone each other to a chisel edge—and the beaver uses them as such, gnawing down trees for its dam and nipping saplings for food. Two glands at the base of its tail contain castoreum, an oil that keeps the fur slick and waterproof. Its tail, a scaly paddle some 16 inches (40 cm) long, serves as rudder, sculling oar, and alarm sounder when slapped on the water. Each hind foot (below) is as broad as a Ping-Pong paddle and fully webbed. Stroking alternately, a beaver can top five miles (8 km) an hour and outswim a diver in racing flippers. With split nails on the two inner toes it oils its fur, combs out lice, and has been seen picking its teeth of splinters.

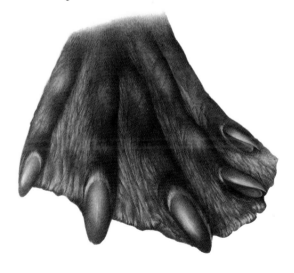

Beaver (*Castor canadensis*)

Its incisors flash a dull orange in their hard coating of enamel. Its scaly tail is so fishlike that 18th-century clerics approved the eating of beaver meat on fast days. Its nostrils are on the sides of its nose, where they can close more easily when it dives. It wins no beauty prizes—yet beaver hats made of the felt from its fur were fashion necessities for centuries.

Beneath coarse guard hairs, the beaver stays warm and dry in an undercoat of fine, dense fur. The demand for this fur lured trappers into the wilds; they opened up a continent but nearly wiped out the beaver. Now its numbers rise.

The beaver gnaws or dies; ever growing incisors, if not worn down, would curve inward and eventually pierce its skull. The animal itself may never stop growing throughout its dozen years. One trapped in Wisconsin weighed 110 pounds (50 kg). A more typical beaver weighs 40 to 60 pounds (18 to 27 kg) and is 3 to 4 feet (91 to 122 cm) long.

To beavers, a pond is security; they seldom venture far from its shore. When bark, twigs, and tubers run short near the edge, they build the dam higher, raise the water level, and bring more food within reach. On land, they are slow, clumsy prey for a host of foes. In the water, the kits fall victim to otters, but adults have few fears.

A beaver can submerge for 15 minutes and surface half a mile (800 m) away. Under winter ice it breathes trapped air bubbles, some of them its own exhalations, and dines on an underwater thicket of saplings jabbed into the bottom in autumn. Predators can cross the ice to the beaver's lodge of sticks and mud, usually at mid-pond, but they find it frozen to an impregnable fortress. Winter and summer, beavers dwell secure.

Nobody's home: An abandoned beaver lodge in Minnesota (opposite) falls into disrepair. Plants ashore and water lilies make a comeback and so may the beavers when these favored foods again abound.

Somebody's home: By a spillway a beaver adds a branch (lower). Beavers sometimes gnaw spillways to relieve the pressure of floodwater. The dam is constantly inspected and repaired.

Lumbermen say beaver ponds drown good timber. But they also provide water to fight forest fires. And conservationists count the beaver an ally, for the dams and ponds it creates help to control flooding while providing a habitat for hundreds of creatures. As nature's lumberjack, a beaver stands on hind legs and tail (below) and bites into a tree. Head cocked, it chews a groove and then another inches below, wrenching out the chip between. In three minutes this tree may topple—and perhaps crush the beaver, for it cannot aim the tree's fall.

High cuts (left) conjure up the 800-pound (363-kg) Castoroides of a million years ago. Probably beavers stood on snow, then cut again after thaw.

OVERLEAF: *Forepaws tucked away in tiny fists, a swimming beaver aims at an entrance to its lodge. Skin flaps close behind its front teeth to keep water out as it swims.*

A hungry beaver strips a bit of bark from an underwater larder. As winter nears, beavers drag shoots and branches into the pond. There in the cold water this food supply stays fresh until spring.

Dry island in a wet world, a beaver lodge rises from a pond that the animals created. A loose array of sticks at the peak allows for ventilation. Residents dry off and sleep on ledges. In this cross-section, one beaver rests on a ledge as others come and go by tunnel. Water from a wet beaver filters through a carpet of shredded fibers that keeps the ledges dry. Here the kits are born—usually four—in April to June, each weighing about a pound (500 g). Within an hour they nurse; in a day, they swim, without a nudge from mother to take to the water. Scrambling out (left), however, is another matter. By autumn they are helping adults store food on the pond bottom for winter. In two years kits are evicted. They then found families of their own.

While a sibling sleeps, three eager beavers line up for lunch (opposite, lower). A lactating female's teats show her sex. But at other times the genders are outwardly identical, their organs of sex, scent, and excretion hidden within a single watertight cloacal opening.

OVERLEAF: Beaver ponds stairstep down the stream that drains broad Cascade Lake in Minnesota. Lodges dot the ponds; dams line their surfaces. Each pond may mark a new generation as young beavers left home and dammed downstream.

Family Muridae

About 1,300 species of mice, hamsters, gerbils, rats, voles, and others worldwide belong to this largest of all the mammal families. Murid ("mouse-like") rodents live on every continent except Australia and Antarctica. In North America, they are found in any area not covered year-round by ice or snow.

Murid rodents mostly eat seeds and vegetable matter. But some are also carnivorous. Muskrats, for example, eat fish, and grasshopper mice eat insects. All of these rodents have 16 teeth. Nearctic members of the family, about 100 species, fall into two broad groups: the pointed-nosed, long-tailed mice and rats; and the blunt-faced, short-tailed voles, lemmings, and muskrats.

The New World murids sometimes have been placed in a separate family, the Cricetidae. In addition, some scientists believe that certain murid groups, such as the voles and their relatives, belong in separate families.

Marsh Rice Rat (*Oryzomys palustris*)

As soon as the rice is planted, rice rats come, like crows to a cornfield, to scratch up their share of the seed. Flooding of the fields doesn't bother them because rice rats are good swimmers.

Making their homes in coastal marshes, riverbanks, canebrakes, and wet meadows, rice rats forage both on land and in water. Their main diet of seeds and succulent green plants is supplemented with snails, insects, small fish, and crustaceans.

The rice rat weaves a grassy nest that may be larger than a basketball. On high ground, the nest is tucked under a log or a tangle of weeds. On flood-prone land, the nest is hung on a cattail or other stem, about a foot (30 cm) above the high-water level. Sometimes the rice rat builds its nest inside a muskrat's lodge.

In the marshy meadows, rice rats cut or bend grasses to form matted feeding platforms. A network of runways connects nest, feeding platform, and water source.

Similar in appearance to Norway and black rats, rice rats are smaller. Total length averages 12 inches (30 cm), and more than half of that is tail.

Rice rats can bear as many as seven litters a year, with three or four young in each. Few live even a year. Hawks, owls, snakes, and carnivores prey heavily on them.

Eastern Harvest Mouse (*Reithrodontomys humulis*)

Ten species of harvest mice, distinguished chiefly by color and size, inhabit the Nearctic Realm. The eastern harvest mouse is found throughout the southeastern United States as far north as Maryland and Ohio.

John James Audubon and John Bachman named this mouse *humulis*, perhaps meaning *humilis*, small. It usually grows no larger than 3 inches (76 mm), plus a 2.5-inch (64-mm) tail.

Closely resembling house mice, harvest mice have hairier tails, pale undersides, and a vertical groove down the front of each upper incisor. Their generic name, *Reithrodontomys*, combines three Greek words and means "groove-toothed mouse."

The harvest mouse lives in meadows and marshes, thickets, and weed-filled ditches. It weaves plant materials into a spherical nest as small as three inches (7 cm) in diameter. The nest, with its tiny entrance hole, often is hung a foot (30 cm) or more up on a plant's stalk, like the nest of a miniature bird. Inside, the nest is lined with thistledown or cattail fuzz. A male and female share this snug home. Several litters are born each year, with two to five babies in each.

Harvest mice work mostly at night, gathering seeds and grain (hence the common name). They climb plant stems, calling to each other in voices so high that humans can barely hear them.

The endangered SALT MARSH HARVEST MOUSE (*R. raviventris*) forages on salty plants and has a high tolerance for salt. Its marshy habitat around San Francisco Bay is jeopardized by urban development.

Oldfield Mouse (*Peromyscus polionotus*)

Six inches (152 mm) from nose to tail tip, the oldfield mouse ranks as the smallest of the Nearctic's 27 species of *Peromyscus,* the "pouched little mouse."

Though their color varies widely from nearly white to almost black, all species have white or pale feet and undersides. They carry food to storage in pouches inside their cheeks.

These mice are found in every sort of habitat—forest and field, desert and mountain. Species with different food preferences may share the same area.

Oldfield's name reflects its most common habitat: abandoned farm fields. As forests grow back, these mice depart and other mice move in. Oldfields also like sandy beaches. Shore populations, paler than their inland cousins, are harder for predators to spot on the sand. But several beach-dwelling races have been nearly wiped out by human disturbance of their habitats.

A little mound of earth marks an oldfield's burrow entrance, which is closed most of the day. Oldfields cache food—seeds, nuts, and berries—in their burrows for winter eating. Like all *Peromyscus* species, they leave food and droppings in their nests. When a nest becomes soiled, the mouse abandons it and builds a new one.

Deer Mouse (*Peromyscus maniculatus*)

Most widely distributed of all *Peromyscus* species—and all North American rodents—this mouse inhabits every sort of dry land from Alaska through most of Canada and the United States, except the Southeast. Its color varies from pale gray to reddish brown.

Deer mice are difficult to distinguish from white-footed mice (*P. leucopus,* next page). Even their common names are sometimes used interchangeably, since both have white feet and the white underparts that remind us of the coloration of white-tailed deer.

Deer mice average a total length of about 7 inches (178 mm) and weigh about an ounce (28 g).

Their cup-shaped nests are built in burrows abandoned by other animals or in hollow logs, stumps, tree root cavities—any sheltered spot. They may roof over a bird's or squirrel's old nest. Often they move into a cabin or shed, or even the pocket of a forgotten coat.

Deer mice eat almost anything: seeds, nuts, berries, insects, worms, spiders, bird eggs, young birds, dead mice. As much as a pint of vegetable food may be stored for winter. Deer mice do not hibernate, but in winter some may sleep huddled in a heap to keep warm.

In warmer areas, deer mice may breed all year, producing as many as eight litters. There are four young in the average litter.

Unlike most rodents, females may permit their mates to share such baby-sitting chores as retrieving strays.

Born blind and hairless, deer mice gain fur and sight within two weeks, become independent in six.

White-footed Mouse *(Peromyscus leucopus)*

Look-alikes to people, deer mice and white-footed mice can certainly tell each other apart. They do not interbreed and cannot produce hybrids. Although the two species coexist in many areas, *P. leucopus* prefers brushy and wooded habitats. Its range extends over most of the central and eastern United States, except the extreme Southeast.

An average whitefoot's total length is about 7 inches (178 mm), the same as the deer mouse's.

Whitefoot's reddish-brown color varies from cinnamon to tawny.

P. leucopus—white-foot—is also called the wood gnome or woodland white-footed mouse. Because of its sylvan preferences, it usually nests high in trees. But it may choose a stone wall, a haystack, or even an abandoned beehive. The mouse moves to a new nest when the old one becomes soiled. As winter approaches, it stocks food caches close by its nest.

One of the most abundant mammals, white-footed mice average three or four per acre (0.4 ha), but some years the number is 20. Breeding may continue all year. There are four in the average litter.

A white-footed mouse's fastidious after-dinner grooming is marked by behind-the-ears thoroughness. It usually begins cleaning and combing at the nose. It may turn acrobat to scrub its back and end at the tip of its tail (above).

Northern Pygmy Mouse (*Baiomys taylori*)

Cactus Mouse (*Peromyscus eremicus*)

Sandy gray fur matches the sandy habitat of the cactus mouse, found in low, hot Southwest deserts from California to Texas and throughout northern Mexico.

Its grassy, fluff-lined nest is built in burrows or hidden amid rocks, clumps of cactus, or any secluded ground-level spot, such as an abandoned building. Cactus mice will also scramble high up trees and bushes in search of food: insects, seeds, nuts, and perhaps green vegetation. Some aestivate, conserving water and stored food.

Also called "desert deermouse," the cactus mouse sports the white underparts and white feet typical of all *Peromyscus* species, with a faintly bicolored tail.

The mouse's tail accounts for more than half its average length of 7.5 inches (191 mm). Large eyes and ears provide extra-sensitive sight and hearing, vital for a nocturnal creature sought by owls, snakes, skunks, and coyotes. Fecundity preserves the species: three or four litters a year, each with one to four young.

Compared to a deer mouse's home range of one or two acres, a pygmy mouse's range is minuscule—less than a fifth of an acre (0.08 ha). Within this tiny world lives North America's smallest rodent. It is about 4.2 inches (107 mm) long—a third or more is slender, hairy tail—and it weighs about one-third of an ounce (8.5 g).

The pygmy mouse lives under a protective canopy of dense grass, building its nest on the ground or in slight depressions under logs or fallen cactus. Pygmy mice are tolerant of one another's presence, but populations seldom exceed eight per acre. Snakes are their major predators; coyotes, skunks, hawks, and owls also take a toll.

Pygmy mice breed year-round, producing litters of one to five after a gestation period of about 20 days. Newborn pygmy mice weigh only one gram, the weight of a shelled peanut. They are weaned in three weeks and are ready to breed in four weeks.

Grass seeds and leaves and other vegetation constitute their main diet, but pygmy mice also climb prickly pear cactuses to nibble the dark red fruit. Trapped mice provide sure evidence: chins and bellies stained red with fruit juice.

B. taylori (range map opposite) is sometimes called the northern pygmy mouse to distinguish it from a neotropical species.

Golden Mouse *(Ochrotomys nuttalli)*

The golden mouse can climb a tree like a squirrel and hang by its tail like a monkey. In its garb of molten gold accented by white underparts, it ranks as the handsomest of mice. Of medium size, it averages about 3 inches (76 mm) of head and body, with a tail of similar length. It is the only member of its genus, though some scientists have placed it in *Peromyscus* in the past.

Throughout their range, golden mice are found in dense woodlands, swampy thickets, tangled vines and underbrush, and particularly where honeysuckle

◻ Golden Mouse
◻ Pygmy Mouse

and greenbrier thrive. Primarily nocturnal and arboreal, the mice gather plant seeds and carry them in cheek pouches to feeding platforms. Woven of grass and shredded bark, they perch high among vines and branches. Scatterings of seed hulls surround a platform. There may be several for each main nest, a globular structure about seven inches (18 cm) in diameter. A bird's old nest may be remodeled and roofed over. Several adults usually occupy each nest. But when a female bears a litter— two or three young—all other adults, including her mate, move out. Golden mice breed all year in warm areas.

Western Grasshopper Mouse (*Onychomys torridus*)

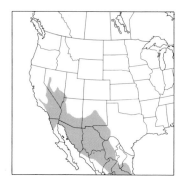

In the dark desert night, a hungry predator stalks its prey. Then, with a rush, it seizes a pocket mouse from behind and slays it, weasel fashion, with a bite to the head. As the predator eats, it may let out a howl.

But this predator is neither wolf nor weasel. It is a mouse, no more than 6.2 inches (157 mm) in body length with a stumpy tail of about 2.4 inches (61 mm). When it howls, it rears up on its hind feet and lifts its head. The howl—which may be a territorial warning—is long and high-pitched.

Of the range shown on the map, *O. torridus* occupies the west and the related *O. arenicola* the east and south. The shorter-tailed *O. leucogaster* is found generally in the north. These three species are among the few carnivorous rodents of the Nearctic. Both have stomachs adapted to meat: lizards, grasshoppers (as above), scorpions, crayfish, other grasshopper mice. Since victims' bodies are mostly water, the mice can go without drinking—unless prey is scarce and the meat eaters must turn to seeds and vegetation.

Grasshopper mice live in close family pairs. Male and female help prepare a nest in another rodent's appropriated burrow—or dig their own. Both care for the young. There are at least two litters a year, each consisting of two to six young. Grasshopper mice are never very abundant. A territory, fiercely defended, averages about seven acres (2.8 ha).

Hispid Cotton Rat *(Sigmodon hispidus)*

As colonists cleared and planted America's southland, they created an ideal habitat for a long-haired, grizzled rat of medium size. As it became more and more populous in the cotton fields, it earned the name cotton rat—in a rough, or hispid, coat. *S. hispidus,* most widespread of the eight cotton-rat species, is also usually the most abundant mammal wherever it is found. Essentially a tropical species, it is spreading northward in the Midwest. One study shows an expansion of 200 miles (322 km) in 40 years.

Populations average 10 or 12 per acre, but a record 513 rats were counted on one acre. Density depends largely on rainfall. Dry years deprive the rats of high-grass cover, and they will breed less often. Hawks, owls, coyotes, foxes, and many other predators feed on them. A cotton rat's life expectancy is about six months.

But high fecundity maintains the high populations. Litters average five or six young and may number 12. The young can breed at six weeks. Females can mate within hours after giving birth. Theoretically, one female could have 144 young in a year.

Cotton rats are dark in the east, paler in the west. Black or brown hair is grizzled with buff or gray. The average length is 11 inches (279 mm), including a 4.3-inch (109-mm), sparsely haired tail.

They build networks of broad runways, with nests in sheltered surface spots or in shallow burrows. They will eat bobwhite eggs and chicks, but feed mostly on vegetation. They can do considerable damage to sugarcane, cotton, and other crops, including young tree plantings.

Eastern Woodrat *(Neotoma floridana)*

This rat by any other name might be prized by humans for its meat. Those who have tried it say it's as tasty as squirrel. Indians relished the plump rodent, which weighs 7 to 13 ounces (198 to 369 g) and measures up to 17 inches (43 cm); almost half of it is tail.

It has soft, well-groomed brownish gray fur with white feet and underparts. A vegetarian, it forages at night for buds, seeds, nuts, berries, grasses, and herbs. It drinks sparingly. Large eyes, oversize ears, and long, sensitive whiskers equip the rat for its nocturnal life.

A wide-ranging species (map on page 153), it is at home in southern woodlands, on slopes of the Appalachians, and on the plains as far west as Colorado. (Most other *Neotoma* species favor western habitats.) Sticks, leaves, and almost anything else make up this pack rat's bulky nest. The inside, typically lined with grass and shreds of paper or bark, is soft and cup-shaped. In such an inner nest two to four young are cradled. One resourceful mother made her nest of pilfered steel wool.

Like others of its kind, the eastern woodrat often drops what it's carrying to pick up a shiny object—a practice that has earned it the sobriquet "trade rat." Woodsmen tell a tall tale about a trade rat that steals a 50-cent piece and leaves two quarters.

White-throated Woodrat *(Neotoma albigula)*

In the arid Southwest, pack rat middens—tarry deposits of fecal wastes and plant material—yield clues to the distant past. They indicate that some 54,000 years ago pack rats gathered the seeds and needles of plants that then were common. Sheltered in dry, rocky places, some middens also suggest continuous use of habitats by countless generations.

Now, as in prehistoric times, *N. albigula* scurries in the night for objects to add to the heap of dross and glitter that is its nest. Mixed in that den of sticks—some as heavy as the average 7-ounce (198-g) resident—may be sections of cactus, cow chips, stones, beverage cans, even a child's toy. (The purloiner above, tail held level like a balancing pole, treads gingerly

among spiny cholla and prickly pear.) Designed for comfort despite its rubbish-pile appearance, a den provides relief from desert heat.

Cactus building blocks are used for food and shelter. Sharp spines protect a cozy nest of grass and fibers woven deep within the den. Risking impalement, nursing young—usually twins—cling to teats when the mother moves.

Woodrats will flee from a rattler and may survive its venomous bite. In the open they are fair game for owls, coyotes, and bobcats. They host in their nests blood-sucking "kissing bugs," so named for biting sleeping campers on the lips.

A pack rat's brush-and-cactus den is weathertight. Among sticks and stones may glitter cans, foil, and perhaps even a gold nugget.

Bushy-tailed Woodrat *(Neotoma cinerea)*

Denounced as a sleep robber and a thief, this "mountain rat" has plagued cabin dwellers from Alaska to Arizona. It pulls batting from mattresses, shreds clothing, and will steal everything from a watch to a bottle cap. A large rat that weighs a pound (454 g) or more, it scampers over shelves and rafters, leaving a urine trail and a strong musk odor. Almost half its 12- to 18-inch (30- to 46-cm) length is tail.

The rat's booty may be added to a bushel (32 l) or more of sticks, bones, rags, and foliage. In this jumble is concealed a nest of soft grasses and bark or fur. Caches of pine nuts, fir twigs, aspen leaves, and other vegetation sustain the rodent through the winter. It does not hibernate.

Besides nesting in abandoned mines and cabins, this rat will live in caves, cliff crevices, among tree roots—almost anywhere from sea level to mountain slopes 14,000 feet (4,267 m) high.

Young—usually three or four—are born each spring. A second litter often follows. Snakes, owls, weasels, and coyotes help to keep bushytails in check.

☐ Bushy-tailed Woodrat
☐ Eastern Woodrat
☐ White-throated Woodrat

Southern Red-backed Vole *(Clethrionomys gapperi)*

Cool, damp forests, white cedar swamps, mountain ridges, and ferny glades are the habitat of this shaggy, stocky vole. Redbacks are abundant in such areas throughout most of Canada, the northern United States, and southward in the Appalachian and Rocky Mountains.

This vole is distinguished by a broad, bright chestnut stripe from forehead to base of tufted tail. In many areas, the redback may be gray with a brown or sooty-gray dorsal stripe. Total length averages 5 or 6 inches (127 to 152 mm).

Active year-round—more often in darkness but also in daylight—redbacks use other rodents' runways. Nests are small spheres or simple platforms woven of grass, leaves, and moss and tucked under logs or tree roots. They eat greens, berries, seeds, and occasionally the carcasses of mice.

Redbacks are highly irritable and nervous. A captive may faint or even die when handled.

Redbacks breed about eight months a year, producing several litters of three to eight young—and they can become parents within four months. But mossy woodlands are also the home of many predators—from shrews and bears to hawks and snakes. Few redbacks survive more than a year.

Twenty-seven species of Nearctic rodents are called voles. (See species list, page 396.) *Vole* comes from a Scandinavian word that means "meadow."

Red Tree Vole (*Arborimus longicaudus*)

So seldom seen that they once were considered rare, these voles live in forests of spruce, hemlock, and fir. Females seldom come down to earth, and they probably outnumber males two to one. The males seem to live mostly on the ground, making nests in shallow burrows or brush piles. How do red tree voles get together to mate? No one knows for sure, but it seems likely that the males move into the trees, build temporary nests, then leave.

The females' nests grow larger with each generation. A nest, perhaps half a bushel (16 l) in size, usually hangs 15 to 50 feet (4.5 to 15 m) above the ground. There, at any time during the year, are born two or three young, naked and blind. They will grow to be about

7 inches (178 mm) long; more than half of that is tail.

Loggers send the voles' forest worlds crashing to earth. Bobcats, raccoons, and owls hunt them, as do jays, which team up to get voles. While one jay tears up the nest, another stays on the ground to poke through fallen bits of nest for helpless young.

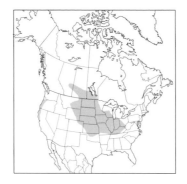

☐ Prairie Vole
☐ Red Tree Vole

Prairie Vole (*Microtus ochrogaster*)

At dawn and at dusk, prairie voles run along paths to feeding grounds of roots, flowers, seeds, grasses, and sedges. (They also dine, as above, on crops.) Voles make paths, called runways, by clipping the grass closely and tramping a path. It is worn into a rut by the patter of little feet. Tall grass—or snow in winter—canopies the trails and helps hide the voles from hawks and owls. But coyotes and weasels sniff them out.

M. ochrogaster—its name refers to its yellow-tinged belly—builds nests above and below ground. But prairie voles spend most of their

time working in their runways. Food is stored in underground chambers, where relatives share the bounty. One such cache held two gallons (7.5 l) of soft plant parts. Each day a prairie vole eats its weight in food, including meat. Mothers fight to save young from cannibalistic adults.

Litters of about four young each are produced several times a year. Females mature within a month. By the time they attain adult size—5 to 6 inches (127 to 152 mm), including tail—they can have mated twice.

Meadow Vole *(Microtus pennsylvanicus)*

A prolific captive female gave birth to 17 litters in a year. She was, in effect, a perpetual breeder, for each gestation lasted three weeks. Had all her offspring plus those of succeeding generations that year been as prolific, the laboratory would have been filled with more than a million voles.

Exploding populations of such magnitude do not occur in the wild, where litters average only three a year—each with five or six young—and where voles fall prey to a legion of animals, from birds to bears—and even fish. A pickerel or lake trout will snap up a swimming vole. One study showed the average life span of a meadow vole was less than a month. Few lived beyond a year.

During peak population periods, numbers per acre (0.4 ha) soar into the hundreds—and, by one record, up to 12,000. A plague of voles can strip an alfalfa field. Voles graze like cattle and horses, with teeth similarly adapted for grinding. And they eat night and day, often sitting on their haunches (like the one above), eating food held in their forepaws.

Its namesake state is only one of many homes for the meadow vole. Most widespread of its genus, the rodent is found from the Carolinas to Alaska. It is 3.5 to 5 inches (89 to 127 mm) long, with a tail half that. It weighs 1 to 2.5 ounces (28 to 71 g).

The vole's grass-roofed runways weave through almost every meadow in Canada, where it is valued as a major food for furbearers. An efficient little machine for converting grass to meat, it has been called "one of those crucial forms upon which the balance of nature rests."

Woodland Vole *(Microtus pinetorum)*

Beady eyes, small ears, blunt head, and short legs mark this "mole mouse" as a burrower. Soil falls from its velvety auburn coat as smoothly as water off a duck's back. This pudgy rodent—5 inches (127 mm) long, with an inch-long (25-mm) tail—is active night and day. It digs shallow tunnels, leaving in its wake ridges that look like the work of a mole. The vole has been clocked digging at 15 inches (38 cm) per minute.

Invading gardens to feed on seedlings and tubers, it can lay waste to a potato crop—and so it has been dubbed potato mouse. Apple mouse is another nickname. Its teeth gnaw bark and roots, sometimes killing fruit trees. Caught in a trap, it may become a meal for one of its own kind.

Infrequently it ventures above ground to forage in leaves (as above). There it is fair game for hawks, owls, skunks, cats, raccoons, and other predators. Snakes and shrews readily enter its burrows, posing a more serious threat. They raid leafy nests where three or four times a year a female vole whelps triplets, sometimes more. Born naked and blind, the young mature in about six weeks. Some females and their litters may share a communal nest.

Although the original specimen was from Georgia pines—hence *pinetorum*—these voles seem more at home amid hardwoods.

Yellow-cheeked Vole *(Microtus xanthognathus)*

Phantom of the far north, this rarely seen vole literally may be here today and gone tomorrow. Irruptions happen only every 20 years or so, say Yukon villagers. Their dogs gorge themselves sick on voles during peak populations. Colonies appear—and disappear—in a variety of habitats, from grassy flats to spruce-clad slopes.

Burned-over or logged woodlands carpeted with sphagnum make choice homesites. Along runways radiating from mounded entries of burrows, chirping males chase females in the spring. Three weeks after mating, litters of seven to eleven are born in sedge-lined nests.

The rat-size yellowcheek weighs 4 to 6 ounces (113 to 170 g) and measures up to 9 inches (229 mm) from the tip of its short tail to its bright muzzle, which is more brownish orange than yellow.

The YELLOWNOSE VOLE, also known as the ROCK VOLE (*M. chrotorrhinus*), has similar markings but is smaller. It ranges from the Great Smoky Mountains to Labrador.

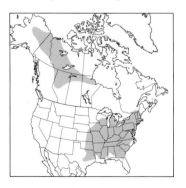

☐ Yellow-cheeked Vole
☐ Woodland Vole

Muskrat (*Ondatra zibethicus*)

Where cattails and clumps of marsh grass grow, the muskrat builds its domed lodge. Largest member of the family Muridae, it thrives in freshwater marshes and along streams almost everywhere in North America from Alaska and Canada to the Gulf of Mexico.

Mudcat, muskbeaver, musquash: The furbearer has many names. "Muskrat" describes the animal's ratlike appearance and the penetrating odor of the musk glands' secretion. This the male deposits along trails at mating time to warn other males of his presence; it also tells females he is available. They too secrete musk.

Though the muskrat builds lodges and complex runways, sometimes tunneling under a stream from bank to bank, it is not a close relative of the beaver. It is not a rat, but more nearly a large, amphibious field mouse. It measures 16 to 25 inches (41 to 64 cm) and weighs 3 pounds (1.4 kg). The glossy brown fur waterproofs the animal and holds body heat in the coldest weather. Trappers since the days of the early explorers have coveted the dense, glistening fur.

Muskrats seldom stray far from water. Excellent swimmers and divers, they paddle with hind feet that are partially webbed. Like beavers, muskrats can forage up to 15 minutes underwater. Submerged, they gnaw tender roots and stems, carry them to the surface, and dine in peace on a raft of reeds. In Louisiana, the remains of mussels, clams, and crayfish have been found on platforms where muskrats feed.

Several times a year the male searches relentlessly for a mate, often waging fierce battles with other males. About a month after mating, the female gives birth to three to eight young. She may have two to eight litters a year. She mates soon after giving birth. While the young are suckling, the male beds down in a separate nest on the opposite side of the lodge. As soon as the young reach sexual maturity—six weeks or so—they are driven out to make room for other litters.

Young muskrats fall prey to mink, snapping turtles, foxes, raccoons, otters, and bobcats. Hawks are threats during the day; owls by night. Confronted, muskrats will stand and fight. A Canadian naturalist watched a muskrat battle a fox five times its weight. As the fox circled, the rodent jumped and bit its tormentor's muzzle. The bleeding fox soon gave up and slunk away.

Despite life's dangers, including reclamation projects that diminish wetlands, prolific muskrats remain abundant. Trappers take nearly 10 million of them every year.

157

Snug and dry in a nest of marsh reeds, the muskrat sleeps and eats. As winter approaches, this semiaquatic rodent erects a new house or repairs an old one. On a mound of mud in shallow water it heaps pondweeds, cattail stalks, and other plant material. Then the animal dives to the bottom and burrows into the mound. Gnawing with sharp incisors, it carves several underwater plunge holes and hollows out living space above the waterline.

Snow-covered muskrat lodges on the Minnesota River (right) rise about three feet (1 m) above the surface. Their tenants swim under the ice to feed on plants and fish. When food is scarce, they stay home and nibble the living room walls. Here the young are born, usually in spring and summer. Or sometimes the birthplace is a bankside nest or even the middle of a pond on a feeding raft of rushes. One muskrat was seen giving birth to seven young on such a raft. The mother grasped the babies in turn and carefully held each kit's head above water as she paddled safely to her lodge.

Collared Lemming *(Dicrostonyx groenlandicus)*

Tundra dwellers, collared lemmings wear luxuriant fur coats of brown, chestnut, and gray. After their autumn molt, they look like fluffy cotton balls. Among rodents, only collared lemmings don white winter coats. Early Eskimos believed the little animals swirled down from the sky with snowflakes.

At home under snow or tundra sod, these tireless diggers develop enlarged front claws as winter nears. Numbering in the millions, lemmings nibble away much of the tundra's fragile plant life. But the many holes they dig allow the sun's heat to penetrate the upper permafrost, thus reducing the size of the polar desert.

A female lemming may give birth several times a year to as many as 11 kits. As adults they measure 4 to 7 inches (102 to 178 mm) and weigh about 2 ounces (57 g).

D. groenlandicus lives only in Alaska, northern Canada, and a strip of Greenland. Lemming numbers fluctuate in cycles with the weather and the food supply—from 50 animals an acre (0.4 ha) to one in 10 acres (4 ha). Population peaks send them on short migrations to find food.

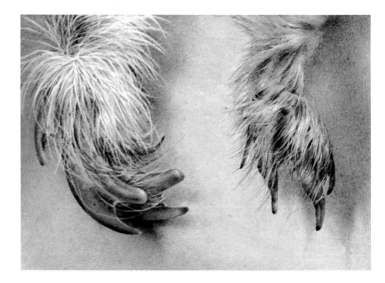

Winterized forefeet equip this lemming for burrowing in snow. Two middle claws lengthen, and claw pads bulge into horny digging tools. In spring the foot returns to its former shape (right).

Southern Bog Lemming (*Synaptomys cooperi*)

In a spongy bog or woodland glade thick with leaf mold, the southern bog lemming makes its home. Night and day it scurries about, mincing succulent grasses. In central Kentucky, dense stands of

clippings line runways and flank entrances to the feeding stations. Underground or in a grass tussock, the female annually bears two or three litters of up to eight young, though one captive lemming bore six litters in 22 weeks. Adults, brownish gray and short-tailed, are 4 to 5.5 inches (102 to 140 mm) long and weigh from one-half to one and a half ounces (14 to 43 g).

The slightly larger NORTHERN BOG LEMMING (*S. borealis*) ranges across meadows and bogs from Labrador to Alaska, south to parts of the northern United States.

- ☐ Brown Lemming
- ☐ Southern Bog Lemming

bluegrass also provide favored homesites. Zoologists use tooth and skull data and *S. cooperi*'s very short tail to distinguish it from the meadow vole.

Bog lemmings dig short tunnels with side chambers for resting, feeding, and storing food. Grass

Brown Lemming (*Lemmus sibiricus*)

For many animals on the tundra—wolves, foxes, owls, hawks, seabirds—the brown lemming is a vital link in nature's food chain. The lemmings in turn feed on tender grasses and sedges.

Every two to five years, the slopes are eaten bare, and lemming numbers decline. Some scientists believe the stress of overcrowding causes hormonal changes that can prevent breeding.

In "lemming years," numbers peak and some animals emigrate to survive. They move into towns and swim across bays and lakes. Carcasses have been sighted on the

frozen sea, 35 miles (56 km) from land. Legends of lemming mass suicide at sea, however, stem from the migrations of a closely related Scandinavian species.

Breeding may go on all winter beneath the snow. Studies have shown that females can become pregnant at the age of two weeks. One pair produced eight litters in 167 days. The male then died.

Brown lemmings measure about 5 to 6.5 inches (127 to 165 mm) and weigh 2.5 to 4 ounces (71 to 113 g). Tawny in winter, they may change to gray in summer.

Family Zapodidae

The "bigfoot" mice called zapodids are distinguished by three-color coats, large hind feet and legs, and tails longer than their bodies. Their multicolored fur makes fine camouflage in meadows and woodlands. They get even better protection from their extra-long tails and hind limbs, for jumping mice can elude predators in swift, six-foot-long (1.8-m) bounds. Their oversize hind feet, splayed outward from the body, provide powerful propulsion. Their long tails give vital balance and can be used to steer and brake as they try to get away from enemies.

These defenses do not guarantee safety. Weasels catch the mice. Their young are dug from their nests by skunks. The first woodland jumping mouse (*Napaeozapus insignis*) ever recorded in Virginia was a victim. It was found in the stomach of a rattlesnake.

Meadow jumping mice of the genus *Zapus* resemble woodland jumping mice. But there is a noticeable difference: The woodland mice have brown fur on their backs, orange along the sides, and white underneath, with a white tip on the tail. The meadow mice are black, yellow, and white. They do not have white tail-tips.

Meadow Jumping Mouse (*Zapus hudsonius*)

Meadow jumping mice inhabit low, moist meadows, marshes, or stream banks, feeding on seeds and berries, fungi, and insects. Mainly nocturnal, they are sometimes seen at dawn or dusk or on overcast days—and may be mistaken for frogs as they pass by in little low hops. Startled, the jumping mouse may first freeze. If its camouflage

doesn't foil the hunting owl or weasel, the mouse catapults away in great, zigzagging bounds. If, in a scuffle, it breaks its tail, it may be doomed to turning involuntary somersaults.

The three species of *Zapus* live only in North America. The most widespread is *Z. hudsonius*. The others are the western jumping mouse, *Z. princeps,* and the Pacific jumping mouse, *Z. trinotatus.*

Z. hudsonius is smaller than its kin, weighing about half an ounce (14 g) and measuring about 7.9 inches (200 mm), including its tapering tail. Its hind feet are more than an inch long (2.5 cm)—a third the length of body and head.

Summer nests are made on the ground or in shallow burrows. Up to three litters, each averaging five young, may be born annually.

Hibernating nests, dug in higher, drier ground, may be as deep as three feet (1 m). Alone or in closely huddled pairs, jumping mice spend more than half their lives in hibernation.

Family Erethizontidae

New World porcupines include eleven species. Ten are found from southern Mexico through the northern half of South America. The eleventh is the porcupine that lives in most of the United States and Canada.

With thickset bodies and short legs, porcupines move at a slow, ungainly gait, foraging in woods and along stream banks. Their feet are modified for tree climbing: wide soles, curved claws, and, on each hind foot, a movable pad instead of a first digit.

Within the porcupine's woolly fur lie hairs modified into loosely attached quills, each controlled by a separate muscle. These quills, some as long as five inches (13 cm), taper to a stiff point covered with barbs (below). The quills are hollow, providing buoyancy that makes the chunky porcupine an able swimmer when necessary. Coarse guard hairs, eight to ten inches (20 to 25 cm) long, give porcupines their shaggy look.

Porcupine (*Erethizon dorsatum*)

The porcupine's quills inspired both its generic name (Greek for "irritate") and its common name (Latin for "quill pig"). They are not pigs, but they do indeed have quills: About 30,000 cover head, back, and tail.

Porcupines don't throw their quills, although it may look as if they do. A threatened porcupine raises its quills, turns its back, and ducks its unprotected face. Feet stomping, teeth chattering, tail lashing (with a few loose quills flying off), it presents a frightening sight—except to a fisher (page 255), its major predator. But another, such as a lynx, may get hundreds of quills embedded in paw or jaw. The quills detach easily from the porcupine's skin, and new ones soon will grow in. The predator cannot pull out the barbed quills, which may work in deeper until they lethally enter a vital organ.

Only beavers are bigger than porcupines among North American rodents. The porcupine averages 36 inches (91 cm)—including a 6-inch (15-cm) tail—in length and weighs about 15 pounds (7 kg). It sleeps all day in crevices, logs, or tree nests. At night it wanders in search of food. Any vegetation will do, but the inner bark of trees is preferred. Porcupines mate in late fall. In spring, one baby is born. Its soft, moist quills quickly dry and stiffen. It may live 10 years or more.

An only child gets mother's full attention—but not for long. This six-week-old (opposite), still wearing the dark fur of infancy, will be independent in a few more weeks. Well developed at birth, porcupine babies can walk and raise their quills within hours, and begin learning to climb trees in a day or two.

Porcupines climb tall trees slowly but skillfully to reach tender twigs and buds—and mistletoe. A bristly tail aids in its tail-first descent.

A porcupine may spend several days in the same tree, gnawing so much bark that the tree may be doomed. More annoying to humans, however, is the porcupine's craving for salt. Anything touched by human perspiration rates as gourmet fare: rifle butts, canoe paddles, camping gear.

Its snubby face and slow, clumsy pace have given the porcupine an undeserved reputation for low intelligence. Keen hearing and sense of smell compensate for its poor sight. Protected by fearsome quills, the placid porcupine follows a lifestyle of peaceful solitude.

Whales Great & Small

ORDER Cetacea

"THERE IS NO PROPER PLACE for them in a *scala naturae*," wrote zoologist George Gaylord Simpson in 1945 as he struggled to classify the cetaceans. "They may be imagined as extending into a different dimension from any of the surrounding orders. . . ."

Although classified by the ancients as fish, the cetaceans are mammals. They feed their young with mother's milk and they possess, at least in fetal life, traces of a hairy coat.

The cetaceans of the world include 79 species, of which 47 swim in North American waters and tropical portions of the North Pacific Ocean, including waters off Hawaii. About half of the species are known as whales, the others as dolphins or porpoises. It's not important whether one calls a small cetacean a dolphin or a porpoise, although some people distinguish "dolphin" as a cetacean having a distinct snout or beak, while "porpoise" usually means one having a smoothly rounded forehead. The variety of common names applied to cetaceans reflects the bewilderment of early seafarers as they tried to classify these unlikely beasts.

Commercial whale hunters use the name "great whales" for the eight largest and most valuable species, including (in order of bulk) the blue, right, bowhead, fin, sperm, humpback, gray, and sei.

Modern cetaceans, wholly aquatic, probably evolved from land mammals 60 million years ago during Paleocene time. On the evidence of blood chemistry—which represents a sort of genetic-memory storage bank—the land ancestors of the cetaceans were similar to the ancestors of cattle and sheep. Presumably those pioneer predecessors emigrated from the continents, and groups of their descendants evolved into the mysticetes (the baleen cetaceans) and the odontocetes (the toothed cetaceans). Cetaceans must have been preceded by forms that lived along seacoasts and in brackish waters, but no fossils of such forms have been found.

Although the slender, snakelike archaeocetes, the oldest fossil forms, date back 45 million years, their bone structure tells us that they were unmistakably whales. All are now extinct. Nearly as ancient are the oldest odontocetes, some of which later may have given rise to the mysticetes.

All cetaceans "remember" their geologic past in anatomical traces known as vestiges or relics. All retain a few coarse hairs up to the time of birth, while some keep several hundred throughout life. Adults have vestigial hips or slender bones embedded in the flank muscles. Fetal whales have hind-leg buds that disappear before birth, and one in many thousands is born with tiny, partly formed hind legs protruding from its body.

The ancestors of whales entered the shallows of the Paleocene ocean hunting for fish and shellfish. As they ventured farther and farther to sea, they were challenged by a new and unfriendly environment that was chilling, fluid, and three-dimensional. (It was also salty, but that seems not to have bothered them.) How did they adapt? All cetacean peculiarities of form and function are easily understandable if one reflects on these three challenges of sea life.

In the chill waters of the ocean, a large volume-to-surface ratio is an advantage to an animal in maintaining a sufficient body temperature. Through natural selection, whales became large-bodied. The smallest cetacean in North American waters (a newborn harbor porpoise) weighs 12 to 16 pounds (5.5 to 7 kg); the largest (an adult female blue whale) up to 200 tons (181 kg). Even tropical cetaceans are large, for the thermal conductivity of water is more than 20 times that of air, which means that a warm-blooded animal immersed in cooler water loses heat rapidly.

All cetacean species normally give birth to a single well-developed baby, called a calf, fully able to withstand the chill of the sea. And all cetaceans are enveloped in warm blubber—a tissue rich in fat that lies just beneath the skin. In a large whale, the blubber may be two feet (60 cm) thick and represent 45 percent of the body weight. It serves not only as insulation

but also as a food depot to be tapped for energy during the long fasts that many cetaceans undergo, especially during the mating season.

Influenced by fluid pressures in this new environment, through natural selection the typical cetacean body has become fusiform, torpedo-shaped. The hind legs have disappeared; the forelimbs have become flippers with their joints rigidly locked from the shoulder outward; genital organs and nipples have sunk into pockets where they rest

when not in service; and the external ears have become tiny pits in the skin. The ear opening of a full-grown blue whale is about the diameter of a pencil lead. Nostrils—one in odontocetes, two in mysticetes—are now on top of the head, where they enable the cetacean to "blow," or quickly exchange its breath. The nasal passages do not open into the throat as in land mammals but smoothly join the windpipe. Thus, when a cetacean swims with its mouth open, water does not enter its lungs.

At the tip of the tail are the flukes—wide, horizontally flattened vanes of muscle, tendon, and gristle stiffened axially by the tip of the backbone. Powerful up-and-down movements of the afterbody and the flukes furnish the main thrust in swimming. According to one zoologist's calculations, an 85-foot (26-m) blue whale swimming at a top speed of 23 miles (37 km) an hour uses 276 horsepower.

"I once watched an eight-foot [2-m] bottlenosed porpoise being netted off Cape Hatteras," wrote zoologist Remington Kellogg, "and was astonished by the strength in his tail. A foolhardy onlooker tried to stand on the flukes. With an effortless flip, the porpoise tossed him several feet."

I myself have often looked down on dolphins racing beside the bow of a ship and marveled at their easy, careless motion.

The killer whale, largest of the dolphin family, and the playful Dall's porpoise are the speediest cetaceans; they can sprint at more than 30 miles (48 km) an hour.

Whalemen have coined names for characteristic postures and movements of whales. In *sounding*, the whale lifts its tail flukes high in the air and dives almost vertically. In *breaching*, it jumps clear of the water and falls on its side or belly with a great splash. In *lobtailing*, it stands vertically with its head underwater while thrashing the surface into foam with its tail. In *spy-hopping*, it assumes a posture like a floating bottle, lifting its head into the air and searching the horizon.

While most land mammals live in horizontal space, cetaceans live also in vertical space. They move up and down in the sea,

locating food, dodging enemies, and keeping contact with one another. They have acquired organs and skills that enable them to read the vibrations, both sounds and ultrasounds, that flood them in their boundless environment. In turn, cetaceans give off signals of various kinds, some poetically called voices or songs. By emitting signals some whales not only talk to one another but also recognize through echolocation the kinds of fish they are pursuing, the shape of the seafloor, and the

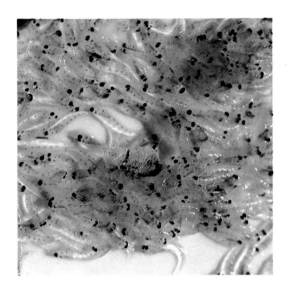

shape and texture of other submarine objects. The ability to echolocate has not been shown in baleen whales, perhaps because typically they passively "graze" plankton and don't chase individual fish and squid.

Cetaceans can vocalize at rates as low as 20 cycles per second, making a sound like the creak of a rusty hinge, or as high as 256,000 cycles. The human ear can pick up sounds no shriller than 20,000 cycles. At full volume, a whale's underwater sounds may carry for hundreds—perhaps thousands—of miles. X-ray studies have shown that in the bottlenose dolphin some sounds are produced within a complex system of nasal sacs rather than within the throat.

Most cetaceans see fairly well both in and

Straining for a living, a baleen whale (opposite) opens cavernous jaws to gulp seawater rich in plankton. Baleen plates sieve the sea soup, retaining shrimplike krill, staple food of these whales.

A movable feast, shoals of krill swarm in the oceans—perhaps 500 million to 5 billion tons of animal protein. A humpback may use its tongue to roll krillballs, downing 1,300 pounds (590 kg) at one meal.

out of water, although their eyes are short-sighted (myopic) in air. Judging from their brain anatomy, they have little or no sense of smell. Their sense of taste has been little studied. Odontocetes lack typical mammalian taste buds, but they do show preference in an aquarium for one kind of chopped fish over another.

Their diving and breath-holding abilities continue to astound us. The body of an unfortunate sperm whale was found tangled in

for two hours before it emerged to breathe.

In the vast, fluid, and trackless medium in which cetaceans live, many species have found advantage, or survival value, in perfecting the behavior trait known as mutual aid. They will instantly help a companion in distress, offering support with their flippers or body, or standing by in what appears to be sympathy. Care-giving behavior resembles human kindness and is one reason why people like cetaceans.

Earning a name and a livelihood, killer whales attack a 60-foot (18-m) blue whale, the world's largest animal species. Hunting communally, a pod of about 30 orcas penned the blue, then moved in to rip its flesh.

a cable 3,720 feet (1,133 m) below the surface. Sperm whales are believed (on the evidence of radio tracking and sonar) to descend to 10,000 feet (3,048 m) or more.

Gunners of a Norwegian whaling station told physiologist P. F. Scholander that a sperm whale can remain below the surface "spontaneously" for at least one hour, while a bottlenose (*Hyperoodon*) can beat its performance. He was told of a harpooned bottlenose that had remained underwater

Off Vancouver Island, British Columbia, a ferry captain saw a vivid demonstration of caregiving by the killer whale, "wolf of the sea" that chases and eats other warm-blooded mammals. Hearing a crunch astern, the captain turned the ferry about. To his dismay he saw a young killer whale, one of a family of four, wallowing in the sea. "The cow and the bull," he said, "cradled the injured calf between them to prevent it from turning upside down. Occasionally the bull would lose its position and the calf would roll over on its side. When this occurred, the slashes caused by our propellor were quite visible."

Another reason for the popular interest in cetaceans is the widely held notion that these animals are intelligent. Are cetaceans as intelligent as our own species? True, the sperm whale has the largest brain—up to 20.2 pounds (9 kg)—of any animal on earth. In the laboratory, a dolphin has been trained to crudely imitate the speech of its trainer.

Or compare the ratio of "new brain" (neocortical) tissue to whole brain tissue. It is low in primitive mammals and high in recently evolved ones. Percentages of new brain tissue are for the common dolphin 97.8, man 95.9, dog 84.2, rabbit 56.0, and hedgehog 32.4. Score again for the dolphin!

However, the original question *should not have been asked.* The big, convoluted brains of the dolphin and the human evolved in distinctly separate worlds. The one organ now

Consider comparative brain weights. Remove the blubber coat that makes up about 40 percent of a bottlenose dolphin's weight; the brain then represents 2.0 percent of its body. The brain of a naked man represents only 1.7 percent. Score for the dolphin!

Then look at the ratio, brain weight to spinal cord weight in the two mammals. (In general, the higher the ratio the greater the intelligence.) It is about 40 to 1 in the dolphin and 50 to 1 in humans. Score for us!

represents a climax of "whaleness," the other of "primateness." Instead of wondering whether dolphins can reason, or worry, or plan ahead as we humans can, we should be content to admire them for their unique minds and bodies—for their vital architecture that allows them to live in a forbidding world of water where our own naked selves could not long survive.

What little is known of the mating of great whales has been learned through tantalizing

Hours later, the assault halts— reason unknown—and the pack leaves the prey. Minus dorsal fin and bleeding from a 6-foot-square (0.6-sq-m) wound in its side, the blue swims feebly, perhaps soon to die.

glimpses, often from afar. During courtship, the whales may leap from the water and fall back with a resounding smash. They may give love pats with the flippers, or teasing strokes with the body surfaces, or gentle nips with the teeth. Mating may take place either as the two swim horizontally near the surface or as they rise vertically from the water in a clumsy embrace, their snouts in the air.

Mating among dolphins, however, has been recorded in great detail by trained observers peering through the windows of oceanariums. As among whales, copulation is brief, lasting only seconds, and is usually repeated. In its quick timing it resembles copulation in cattle and sheep, those ungulates thought to share ancestry with whales.

The gestation period varies with the species. Among mysticetes it ranges from $9\frac{1}{2}$ to 13 months; among odontocetes from 11 to 17 months. The mysticete fetus begins to spurt in growth as it nears full term; the odontocete fetus grows at a steadier rate.

Twin fetuses occur in about eight of every thousand baleen whales. It is unlikely, however, that a whale could successfully nurse two calves to the stage of independence.

One of the first play-by-play accounts of a cetacean birth was written in 1947 by Arthur F. McBride. Standing at a poolside of Marine Studios, in Florida, he had been watching a female bottlenose dolphin, her abdomen greatly swollen, which had withdrawn from her companions. I reconstruct his story:

Time Zero. Her contractions begin; she gives queer barking sounds underwater.

Minute 7. The soft, folded tail flukes of the fetus emerge and slowly relax in the water.

Minute 17. She arches her back violently, tail upward, and expels the 25-pound (11-kg) fetus as far as its flippers.

Minute 23. Out comes the head! The calf falls free; the mother whirls about, breaking the umbilical cord near the calf's body. The placenta is not expelled for several hours.

Minute 24. The calf instinctively swims to the surface and takes its first breath. Pool companions, intensely curious, join the pair.

Parturition in dolphins lasts from 21 minutes to several hours. Normal births are rapid, the calf usually arriving tail first. The mother squirts milk into the funnel formed by the nursing calf's muscular little tongue.

The mysticetes are thought to nurse their young for about 6 to 11 months, the odontocetes for 18 to 25 months. The largest odontocete—the huge sperm whale—evidently nurses for two years. Dutch zoologist E. J. Slijper wrote that cetacean milk tastes like "a mixture of fish, liver, Milk of Magnesia, and oil." It has the thick texture of canned milk, and its fat content—40 to 50 percent—makes it ten times as rich as cow's milk.

Shortages of available food or seasonal periods of fasting continually interrupt cetacean growth. As a result, growth marks like the annual rings of trees are left in the teeth of odontocetes and in the baleen and waxy ear plugs of mysticetes.

Some of the longevity records among cetaceans are: white whale, 30 years; bottlenose dolphin, at least 32; northern bottlenose whale, at least 37; sperm whale, 77; and fin whale, at least 80.

Navigators supreme, all the great whales migrate from cool waters where they feed in summer to warmer breeding waters in winter. How they navigate is unknown, although, like migrating birds, they doubtless use all their body senses. Their seamarks are thought to be wind currents, water currents, a myriad of underwater sounds, contours of the seafloor, water temperatures, and the position of the sun and moon.

The migration of the gray whale of the North Pacific is famous. A typical pregnant female leaves the food-rich waters of the Arctic Ocean in September and swims southward through Bering Strait to the west coast of North America. In late December she passes near San Diego, where each winter nearly a million tourists visit headlands or embark in small "wildlife safari" boats to watch the whales go by. In early January, she gives birth—along with 1,200 other mature females—in certain sheltered lagoons of

western Mexico. In March, the mother and calf turn northwestward on a course still imprecisely known to us. She weans the calf at sea in late July after nursing it for seven months. From early summer to early fall she feeds in the Bering and Chukchi Seas and the Arctic Ocean. Her 12,000-mile (19,000-km) migratory journey has come full circle.

The movements of odontocetes smaller than the sperm whale are poorly known, although it is clear that most dolphins and porpoises move seasonally (and some even daily) in response to changes in the food supply. In summer, pilot whales off the coast of Newfoundland move inshore in synchrony with the migrations of a squid (*Illex*) upon which they feed heavily at that season. The

white whales of polar seas are less specialized feeders. Their movements are dictated mainly by the shifting limits of the pack ice.

While the baleen whales are "grazers," the white whales, like other toothed species, are hunters. They feed on small schooling fish, squid, and swimming crabs. (The killer whale, an exception, also preys on warm-blooded animals.) Worldwide, the odontocetes number about 70 species, of which 11 are dealt with in this book. Some of these travel hundreds of miles up rivers, while others (though not North American species) live landlocked in freshwater systems.

The teeth of odontocetes are typically conical, sharp, and uniform (not distinguished as incisors, canines, and molars). They vary in

Easy riders, acorn barnacles do not harm their host, a humpback, but the burden of freeloaders may build to 1,000 pounds (454 kg). When a migrating whale reaches warmer seas, barnacles may drop off.

number from only two in the narwhal and in certain beaked whales to 242 in the franciscana, or La Plata dolphin. Odontocetes have permanent teeth only. Although milk teeth begin to form in fetal life, they soon disappear in the growing tissue of the gums.

The power of echolocation, well developed in odontocetes, probably less so in mysticetes, may relate to the difference in food-gathering habits. A sense of surroundings is clearly more useful to a hunter than to a grazer. The longer nursing period in toothed cetaceans also relates to food gathering. Evidently the calf needs more time with its mother to learn complex navigational skills.

The odontocete skull is asymmetrical and is telescoped, as if pushed from the front. The two nostrils unite before they reach the surface, then open in a single crescentic blowhole. Within the head, the nostrils flare into oddly shaped chambers or vestibular sacs. These probably function in sound production. They are absent in mysticetes.

Most odontocetes are smaller than mysticetes. The smallest in North America, the harbor porpoise, weighs, as an average adult, 100 to 130 pounds (45 to 59 kg). Odontocete males are nearly always larger than females.

The mysticetes, toothless whales, number ten species worldwide, nine in North American waters. These cetaceans gather food by means of peculiar outgrowths of the upper gums known as baleen (a word both singular and plural). Early American whalers spoke of baleen as "whalebone," although it is not bone. It is a black, white, yellowish, or gray substance textured like fingernail, densely fringed with hairlike bristles on its inner, or tongue, side. The name "mysticetes" incorporates a Greek word for mustache.

Baleen is a series of plates arranged like the shutters of an open Venetian blind. The plates reach their climax development in the bowhead whale, where they may number 360 on each side of the jaw. Some of the plates are 14 feet (4 m) long.

Plankton organisms known as krill are trapped by the baleen plates and are probably rolled into balls by the tongue and currents of water. They then pass into the slender throat and are rhythmically swallowed. Commercial fishermen along the Nova Scotia coast keep a sharp lookout for baleen whales, for both whales and fish (especially herring) are likely to gather where krill is abundant. Thus whales are indicators of fish schools. An astounding biomass of krill, estimated as high as five billion tons, is the greatest potential source of animal protein in the world. (Whales found this out long before we did.)

Baleen whales also differ from toothed species in being exclusively oceanic. They also are large—5 to 200 tons (4.5 to 181 t)—the females being slightly larger than the males. They have paired blowholes, and they have symmetrical skulls in which the bones are telescoped, as though shoved from behind. They are "hairier" than odontocetes, having numerous coarse "whiskers" (vibrissae) around the jaws and on top of the head.

Ocean life, like life on land, presents a variety of hazards. Cetaceans suffer many of the same ailments that distress people. They are attacked by bacterial, fungal, and viral diseases, including pneumonia and tuberculosis; they suffer cancers and stomach ulcers; their bodies harbor parasitic worms. Certain other "fellow travelers" that attach to their skin, teeth, or baleen do no harm. These include diatoms, other algae, protozoans, barnacles, tiny crustaceans called "whale lice," and suckerfish or remoras.

We continually dump contaminants into the sea, threatening marine life. Poisons such as insecticides, weed killers, lead from automobile exhausts, and manufacturing wastes such as cadmium, mercury, and arsenic find their way into the sea. Here they enter the bodies of primary producers, mainly diatoms. Many poisons resist natural decay, so they end up in the bodies of cetaceans—the organisms highest on the food ladder.

Collectors for a Florida aquarium saved a rare, 570-pound (259-kg) pygmy sperm whale found on a beach. The whale died a week later; it had swallowed a large plastic bag of the kind fishermen carry on their boats. Cetaceans make headline news when they run aground in pods numbering 200 or more. The stranded species are nearly always odontocetes. Because echolocation is known to be highly developed in odontocetes but not in mysticetes, some scientists believe that failure of the echolocation system may cause strandings. If a pilot whale, for instance, overshoots its prey and finds itself helpless in shallow water, its frantic calls of distress may draw its

tolerant—up to a point—of the other's presence. Porpoise remains have been found in shark stomachs and vice versa. In one find, a harbor porpoise had choked to death while trying to swallow a four-foot (1.2-m) shark.

People, of course, are the archenemy of cetaceans. Modern technology has precipitated their decline worldwide. A great many of the smaller species drown in commercial fishing nets; others die from pollution, loss of prey, or entanglement in debris.

companions. Soon the shallows are filled with confusing "white noise" signals and exhausted animals. One cause of echolocation failure and stranding is thought to be parasitic worms that invade the brain by way of the nasal sinuses and ear openings.

The theory has been advanced that cetaceans turn to the land in times of distress because they "remember" the safety it provided for their terrestrial ancestors tens of millions of years ago. Although the theory has a certain romantic appeal, it seems unlikely.

The cetaceans have few enemies. Killer whales are known to attack and eat other cetaceans, while sharks prey on dolphins. As a rule, however, sharks and cetaceans seem to maintain a sort of watchful truce. Each is

International efforts to protect the great whales often have been undermined by those nations that derive the most profit from whaling. Abusive commercial whaling, stimulated by the development of the explosive harpoon in the 1860s and the self-contained factory ship in the early 1900s, may have doomed several species. The reported kill of great whales in the peak decade of whaling, 1956-1965, was more than 600,000. By the 1980s the seven great baleen whales had been reduced by nearly 90 percent from perhaps 1.4 million prior to commercial whaling, to around 150,000. Sperm whale numbers also had fallen. The blue and right whales may be below 2 percent of their original numbers, never to recover.   VICTOR B. SCHEFFER

Born eyes open and able to swim, a pilot whale calf gets a jet assist at suckling time. Mother squirts her milk speedily so the calf can surface. Some newborn whales breathe every 30 seconds or so.

OVERLEAF: *Headed for the deep, says the tail of a humpback. One of the imperiled giants, the humpback is acclaimed for its vocalizations. In concert season—winter—song resounds through its watery realm.*

Family Eschrichtiidae

This family consists of a single species, the gray whale of the North Pacific Ocean. Formerly, perhaps as recently as the 18th century, it lived also in the North Atlantic. Gray whale bones of somewhat less than fossil age have been discovered in sediments along the northwestern coast of Europe in England, Sweden, and the Netherlands.

The map shows the gray whale's migration route and the location of major summer populations.

Gray Whale *(Eschrichtius robustus)*

To North Americans, the gray whale is the best known of all the cetaceans, for millions of people have seen it passing southward near the southern California coast in midwinter. People have also observed it in the lagoons of Baja California in Mexico, where it breeds after its 5,000-mile (8,000-km) migration from polar seas. Reduced by hunting in the early 1900s, it was given protection in 1937. Its numbers have now leveled off at about 21,000.

In February 1972, the Mexican government established a new national whale refuge—the world's first—to protect the gray whale nursery in Scammon Lagoon.

The gray whale has been called a "living fossil." Its skull, in particular, is constructed along the general lines of whales that lived millions of years ago.

It differs from other baleen whales in that it often feeds inshore among rocks and kelp; it may even play in the surf.

The gray whale is found in shallow waters on both sides of the North Pacific from the Arctic

Ocean to Mexico. Commercial hunting had drastically reduced the western Pacific, or Korean, stock by the early 1930s. Very few individuals of the original Asian stock are known to survive.

In the eastern Pacific stock, a typical pregnant female leaves Alaskan waters in September. She swims southward at 5 miles (8 km) an hour, travels some 100 miles (160 km) a day, and arrives in Scammon Lagoon in December. There she gives birth. (She will not remate for a year or two.) She and the calf leave the lagoon in early March and swim slowly northward.

About the end of July she weans the seven-month-old calf.

Gray whales migrate singly or in groups of as many as 16. Not all individuals follow the established migration pattern. They may be seen throughout the year in the Gulf of California, off the Farallon Islands west of San Francisco, and off Vancouver Island, British Columbia, and at other points along the coast.

The gray whale evidently does not feed during its southward migration but lives on fat. It loses

about one-fifth of its weight while in winter quarters south of San Francisco. In summer it feeds largely on amphipods near the floor of the ocean.

Male and female grays mature sexually between ages 5 and 11. They mate in late November and early December; the gestation period is 13 months. They reach full size at age 40 and are known to live to at least 69.

Gigi, a 2-ton (1.8-t), 19-foot (5.8-m) yearling gray whale, was captured in Scammon Lagoon in 1971 and successfully held in a San Diego oceanarium for a year. When

she became too costly to feed, she was released in the open sea with a radio transmitter pinned to her dorsal ridge. Navy scientists tracked her position along the coast of southern California for seven weeks, until her signals faded into silence.

The record length for a female is 49 feet (15 m) and weight 35.2 tons (32 t); for a male, 47 feet (14 m) and 27 tons (24.5 t). The newborn calf measures about 16 feet (5 m) long and weighs about 1,500 pounds (680 kg).

Family Balaenopteridae

These furrow-throated whales are called "rorquals," a name derived from an old Norwegian word. Skin furrows running from the point of the chin to the chest expand like accordion pleats when a rorqual opens its jaws in a wide "yawn," increasing its water intake. The rorquals are also called "fin whales." The term, however, is not distinctive, for many species outside this family also have a dorsal fin.

Besides having furrows, the rorquals are distinguished from the right whale by a wider skull, coarser and less flexible baleen, unfused neck vertebrae, and longer, more tapering flippers. The rorquals have lost one finger bone in each flipper, while all other cetacean species retain the primitive five.

Within this family are six of the world's great whales: the blue, fin, sei, Bryde's, minke, and humpback. All six can be seen in North American waters.

Fin Whale (*Balaenoptera physalus*)

During the 19th century, the fin whale, with a population of about 470,000, was the most common baleen whale. Hunting has reduced its numbers to about 60,000.

The fin whale has been called "greyhound of the sea," for powerful muscles can sustain its tapering body at speeds around 23 miles (37 km) an hour through the water and to submarine depths of at least 755 feet (230 m). Unlike the blue and sei whales, the fin whale sometimes leaps clear of the water, falling back with a tremendous splash. For some unknown reason, the right side of the head is paler than the left.

The finback may measure to 76.1 feet (23 m) and weigh 70.8 tons (64.2 t). The newborn is about 21 feet (6 m) and 2.09 tons (1.9 t). Roy Chapman Andrews wrote that an Alaskan 65-foot (20-m) fin whale discharged a 22-foot (6.7-m) fetus as her body was being drawn up the slip of a whaling station.

Fin whales range through all oceans, but are rare in tropical waters or amid pack ice. They sometimes are found singly or in pairs but more often occur in pods of six or seven. Many pods with as many as 50 animals may concentrate in a small area. A pelagic species, it is not often seen near shore.

Many individual finbacks do not migrate in fall to warmer waters but remain in subpolar waters where food is more abundant.

The diet is mainly krill but includes anchovies, capelin, herring, lantern fish, and other small fish, as well as squid.

Blue Whale (*Balaenoptera musculus*)

The largest animal that has ever lived, this bluish-gray whale was feared by open-boat whalers of the 19th century because of its powerful flukes and its speed when wounded—up to 23 miles (37 km) an hour. By the early 1900s, whalers began hunting the animals in fast, steam-powered catcher boats, killing them with fragmentation bombs fired from cannon and butchering them at sea aboard huge floating factories. In 1931 the whalers of the world killed 29,606 blue whales. Blue whales have been protected since 1966 by all those nations adhering to the International Whaling Convention. Some zoologists doubt that the blue whale will ever recover from overhunting. While the North Pacific population may contain more than 2,000 blue whales, there are only a few hundred in the North Atlantic and an estimated 660 in the Southern Hemisphere.

The largest blue whale both measured and weighed (piecemeal) was a female brought to the whaling station at South Georgia in the South Atlantic about 1931. She was 96.8 feet (29.5 m) long and weighed about 196 tons (177.8 t). Whales slightly longer than 100 feet (30 m) have been reported. A 100-foot (30-m) female at the peak

of her feeding season in the Antarctic summer might weigh as much as 215 tons (195 t).

Blue whales of North American waters feed in summer in the Gulf of Alaska, along the Aleutian Islands (rarely in the Bering Sea), and in Davis Strait. In winter they breed off western Mexico and at an unknown place somewhere in warm waters of the North Atlantic or Caribbean Sea. Blue whales usually travel singly or in pairs.

Blues feed almost entirely on shrimplike krill and may eat over a ton at one feeding. Canadian zoologist David E. Sergeant estimates the daily food demand of a baleen whale is 2 to 4 percent of its body weight. Thus, a 200-ton (181-t) blue whale may consume 4 to 8 tons (3.5 to 7 t) of krill a day.

Male and female mature sexually at about ten years of age. Every two or three years, during winter, the female gives birth after a 12-month gestation period. When the calf is weaned at about seven months, it has gained an average of 7 to 8 pounds (3 to 3.5 kg) an hour during its entire suckling period.

Many tales are told of the blue whale's strength and stamina. One harpooned individual towed the steam-whaler *Puma,* with her engines running at half speed astern, for nearly 24 hours.

Sei Whale (*Balaenoptera borealis*)

Norwegian fishermen gave this whale its name because it arrives on their coasts at the same time as the *sei*, or pollack. Smaller and less valuable than its cousins the blue, fin, and humpback, the sei whale was largely ignored by whale hunters until they had depleted the larger species. Along the Pacific coast, the sei ranges from the Aleutian Islands in summer to Mexico in winter, and along the Atlantic coast from Davis Strait in summer to the Caribbean in winter. The whales travel in small groups of two to five but form larger aggregations while at the feeding grounds. In the far north they feed on copepods and other plankton organisms. In the lower latitudes their diet includes small schooling fish—sauries, anchovies, herring, sardines, jack mackerel—as well as krill. The uncommonly fine, soft-textured fibers that fringe the baleen aid in capturing organisms as small as copepods. A 47-foot (14.3-m) sei whale was examined by scientists of Japan's Whales Research Institute. The plankton-straining surface of the whale's baleen amounted to 39 square feet (3.6 sq m). Thus, every time a sei whale squishes a mouthful of seawater through its baleen it is, in effect, casting a net nearly as large as a kingsize bed. One sei whale

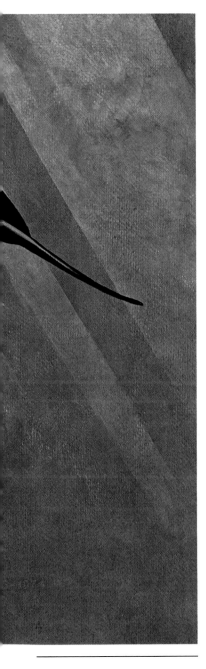

Bryde's Whale *(Balaenoptera edeni)*

Bryde's whale, whose common name honors the captain of a Norwegian whaling vessel, is similar to the sei in appearance, but smaller. It lives in the tropical and warm temperate regions of the Atlantic, Pacific, and Indian Oceans.

Until the middle of the 1900s whalers confused it with the sei. On close inspection, Bryde's can be distinguished by three lengthwise ridges on its snout as compared to the sei whale's one. Its baleen is stiffer and not as fine as that of the sei, and it captures coarser food. Schooling fish—herring, mackerel, anchovies, and others—are major foods. A Bryde's once swallowed a two-foot (61-cm) shark.

Making deeper dives than the shallow-feeding sei, Bryde's also has a higher "blow" as it surfaces. It rises steeply, showing much of its head, rolls its body, and humps its tail without raising the flukes as it begins another dive.

The largest known Bryde's was 45.2 feet (14 m) in length and weighed 17.8 tons (16.1 t). An estimated 30,000 to 60,000 of these whales remain.

that was measured had managed to pack 502 pounds (228 kg) of plankton into its stomach.

The sei has reached a length of 53.8 feet (16 m) and a normal weight of at least 23.8 tons (21.6 t). A very large, pregnant one weighed 41.6 tons (37.7 t). From an original population of more than 200,000, there are now about 36,000 left.

Humpback Whale (*Megaptera novaeangliae*)

The chunky, slow-moving, amiable humpback is known to millions of armchair mariners who read about whales, listen to their recorded voices, or watch them on television. A humpback breaching against a deep blue sky has become a symbolic picture of all great whales.

Its long flippers, up to 14.3 feet (4 m) long, and its head adorned with fleshy knobs and white barnacles are among its characteristic features.

The humpbacks of the world were wastefully hunted until 1966, when they were afforded complete protection. By then, only a few thousand were left, perhaps 7 percent of their original numbers.

The largest known is 51.8 feet long (16 m) and weighs about 51 tons (46.3 t). At birth a calf measures about 15 feet (5 m) and weighs 1.4 tons (1.3 t).

The humpback ranges the eastern North Pacific Ocean from the Chukchi Sea south to southern California during the summer, and from southern California south to the Revillagigedo Islands and Jalisco, Mexico, and also around the Hawaiian Islands, during the winter. In the western North Atlantic Ocean it ranges from Disko Bay in western Greenland south to Massachusetts during the summer, and from Hispaniola and Puerto Rico south to Trinidad during the winter.

Its diet consists mainly of krill, and also small schooling fish. One humpback killed in the North Atlantic had in its stomach six fish-eating birds (cormorants)—and a seventh stuck in its throat.

Charles Jurasz, a high-school science teacher and whale-watcher in southeastern Alaska, has often seen humpbacks creating a "bubble net" to capture herring and krill. "One day," he writes, "I noticed bubbles rising to the surface a short distance away. They formed a perfect circle, and just as the last few bubbles surfaced, there was a tremendous boil of fish right in the middle. Then suddenly a whale surged up from the bottom—mouth opened like a cavern—big enough to swallow me and the skiff."

Northern humpbacks mate between October and March and calve after a gestation period of 12 to 13 months; the calf nurses for about 11 months. An adult "escort"

Minke Whale *(Balaenoptera acutorostrata)*

whale often stays close to a mother-and-calf pair.

Canadian zoologist Peter Beamish once studied a female that had become entangled in a fishing net and was later removed to a holding pen. He temporarily blindfolded the docile whale and allowed her to swim through a maze of metal poles. Although she made clicking noises, these evidently did not help her to echolocate, for she repeatedly hit the poles.

About 25,000 humpback whales remain from an original population of 150,000.

Long ago, a Norwegian whaler named Meincke harpooned one of these whales, mistaking it for a blue. His companions were so amused at his error they began to speak of "Meincke's whale"—now called "minke."

This smallest of the rorquals, distinguished by a white band on each flipper, is fairly common in coastal waters. Along the Pacific coast it is found from the Chukchi Sea south to northern Baja California during the summer and from central California south to central Mexico during the winter. Current estimates put its numbers at about 800,000.

Along the Atlantic coast its range is from Baffin Bay south to Chesapeake Bay during the summer, and from the eastern Gulf of Mexico and northeastern Florida to Puerto Rico and the Virgin Islands during winter. It is most often seen singly or in groups of two or three.

The minke whale feeds mainly on small shoal fish and krill. In the Antarctic it has reached a length of 36.4 feet (11 m); it weighs as much as 9.8 tons (8.9 t). North American minke whales evidently do not exceed 32.8 feet (10 m). The newborn calf is about 8.5 feet (2.6 m) long.

Minkes are more likely to be seen at close range than others of the rorqual family, for they appear to be curious about boats and may swim close to an anchored or slowly moving vessel.

In 1957 a party of British explorers traveling on the sea ice off Antarctica came upon several pools in the ice through which minke, killer, and beaked whales were rising at intervals to breathe. The whales had been trapped by the sudden freezing of a shallow harbor entrance. The men formed a Pat-the-Whale Club for those who touched a whale's snout as it rose.

Natives of the Aleutian Islands used to kill minke whales with stone-tipped harpoons dipped in the juice of a poisonous plant, monkshood *(Aconitum)*. During the Middle Ages, Norwegian hunters killed minke whales with "death arrows" dipped in fluid from a rotten whale. After several days, the victim weakened and could be harpooned. Strangely, its flesh was eaten without ill effects.

Family Balaenidae

Early whale hunters named these the "right" whales because the animals could easily be approached in a small boat and because their harpooned bodies, rich in valuable oil, would float after death.

The arched upper jaws in right whales give the mouth its great capacity to hold water. The floor of the mouth is shaped like a gigantic scoop or spoon. (In balaenopterids, "accordion pleats" on the throat give similar capacity.) The huge skull makes up 25 to 40 percent of the length of the skeleton. Where the seven neck vertebrae join the skull, they are fused into a single bone. The right and left baleen rows do not form a continuous horseshoe shape but are separated by a gap at the snout.

Excepting the pygmy right whale of the Southern Hemisphere, these whales have no dorsal fin. Some scientists put the right whale in a separate genus, *Eubalaena,* and some distinguish the population of the Southern Hemisphere as a separate species, *B. australis.*

Right Whale *(Balaena glacialis)*

What is known of the form and coloration of the now uncommon right whale, or black right whale, of North American waters is based on a few specimens killed under permit for scientific study. The body is black or slaty gray, usually with white patches of irregular shape on the underbody. On the snout in front of the blowholes the right whale bears rough, fleshy bumps that are commonly infested with "whale lice" (actually crustaceans) and barnacles; the whole region is known as the "bonnet."

The record size of a right whale is 60 feet (18 m), and its estimated weight 120 tons (108.9 t). The size of a newborn calf is believed to be less than 19.6 feet (6 m) long. A suckling calf 21.3 feet (6.5 m) long was collected in South Africa. The testes of the full-grown male right whale are enormous. One pair weighed by Japanese scientists tipped the scales at 2,143 pounds (972 kg).

The species is distributed through the temperate oceans of the world. The migratory courses in the Northern Hemisphere are poorly known, though right whales move north in summer and south in the winter.

William A. Watkins and William E. Schevill, at Woods Hole Oceanographic Institution, observed right whales near Cape Cod every year for 20 years. They concluded that the whales congregate in the spring and early summer "singly, in groups of three to eight animals, in adult pairs, and in cow and calf pairs. Occasionally

Bowhead Whale *(Balaena mysticetus)*

loose aggregations are formed with up to 30 animals within a few square kilometers."

Originally numbering 100,000 to 300,000, the species was nearly exterminated by whalers. It was given worldwide protection in 1935. There are only a few hundred left in the North Atlantic and even fewer in the North Pacific. Some 1,500 survive in the Southern Hemisphere.

Right whales feed mainly on the smaller organisms of the krill—copepods about the size of a match head—and on young euphausids. They have been seen grazing near the surface in thickets of reddish krill, seemingly able to stay inside the richest concentrations. Watkins and Schevill once drifted within six feet (1.8 m) of a feeding right whale. They could hear the animal's baleen plates.

The bowhead was named for the resemblance of its strongly curved jaws to the shape of an archer's bow. The species has a huge and powerful head. Eskimos say that a bowhead can break its way up through ice several feet thick. The head represents more than one-third the length of its body. Measuring up to 65 feet (19.8 m), an adult of record length would weigh about 152 tons (137 t). The length of a newborn calf is 13 to 15 feet (4 to 5 m).

The bowhead breeds in summer at the edge of Arctic ice, moving slowly southward in winter with the drifting floes. Its main food source is euphausids and other crustaceans in the krill. Bowheads evidently feed sparingly during spring and fall migrations. As many as 50 may assemble in fall, but the bowhead usually travels singly or in groups of two or three. (The one above swims with belugas.)

The world population of the bowhead, including the Spitsbergen, Davis Strait, Hudson Bay, and Bering Sea stocks, was estimated in 1990 at about 8,000. More recent estimates are slightly higher. But the population is probably little more than 10 percent of its prewhaling size.

In the late 19th century the bowhead was the most valuable of all whales because it yielded copious amounts of oil, used as fuel in household lamps, and the finest baleen, called whalebone and used mainly for corset stays.

Alaskan and Canadian Eskimos, in keeping with their cultural tradition, are still allowed to kill some 50 bowheads each summer; otherwise the species has been protected by all the northern nations since 1935. The Eskimos hunt from skin-covered boats called umiaks, using dart guns or shoulder guns. Either gun hurls a bomb into the whale.

Family Delphinidae

The world's delphinids, including about 40 species, are the most abundant and varied of all cetaceans. Although the larger members of this dolphin and porpoise family are called "whales," nonetheless they are typical delphinids. Unfortunately, quick identification at sea is not easy, for the main distinctive characteristics of the family are skeletal. Delphinids have numerous conical or spatula-shaped teeth in both the upper and lower jaws; they lack throat grooves; and their tail flukes are notched. Most, though not all, have a dorsal fin.

They are mainly fish eaters and oceanic. They feed in upper waters, surfacing to breathe several times a minute. Among bottlenose dolphins that were studied for more than a year, the average dive time per animal was 21.8 seconds.

A game fish, *Coryphaena hippurus,* is also called dolphin, and this has caused some confusion about names. The fish and the mammals are, of course, not related. As to the difference between dolphin and porpoise, the names are often used interchangeably.

Common Dolphin *(Delphinus delphis)*

"Full of fine spirits, they invariably come from the breezy billows to windward. They are the lads that always live before the wind. They are accounted a lucky omen." Thus wrote Herman Melville of the common dolphins. They are indeed a sporting class, leaping in unison as they play at the bow of a moving vessel.

In rare instances a male may reach 8.5 feet (2.6 m) in length and weigh 300 pounds (136 kg). Females are slightly smaller. One of the most widespread and abundant dolphins in the world, it lives in warm and temperate open seas. On the coasts of North

America it ranges from Oregon to Costa Rica and from Newfoundland to the Caribbean. Common dolphins may gather in groups of several thousand.

Two zoologists once watched common dolphins off the coast of California feeding on sardines, anchovies, sauries, small bonito, and squid. Several dolphins drove the prey clear out of the water—and caught it in midair! Their teeth are small, sharp, and recurved, perfectly adapted for catching slippery fish. A United Nations report claims that these dolphins along the California coast eat 300,000 tons (272,000 t) of anchovies each year, whereas commercial fishermen take only 110,000 tons (99,000 t).

Dolphins dive deepest at night—down at least 846 feet (258 m), apparently searching for food within the "deep scattering layer" of plankton and larger organisms.

Russian whale expert A. G. Tomilin traveled on a ship in the Black Sea carrying on deck 90 live common dolphins. All through one night he heard protests from the dolphins that reminded him "of a child's rubber toy or the short whistle of a sand grouse (lasting one or two seconds), less often, the quack of a duck or the yelp of a cat whose tail has been pinched."

Bottlenose Dolphin *(Tursiops truncatus)*

This handsome beast, familiar to millions who visit oceanariums or watch animal programs on television, is symbolic in the public mind of all dolphins. Partly from sympathy for it, the people of the United States demanded the landmark law that conserves other species as well, the Marine Mammal Protection Act of 1972.

Soon after the world's first oceanarium, Marineland, opened at St. Augustine, Florida, its manager began to wonder whether a dolphin could be taught to do tricks to entertain visitors. So in 1949 he hired Adolf Frohn, who trained wild animals for a circus, to educate Flippy, a 200-pound (90-kg) male bottlenose. Within several weeks, Flippy was showing his pleasure at being "in school" by leaping into Adolf's arms!

The size of the bottlenose varies considerably from place to place. The largest on record are a male 12.7 feet (3.9 m) long, from the Netherlands, and a female 10.6 feet (3.2 m), from the Bay of Biscay. A Dutch specimen 9.8 feet (3 m) long was said to weigh 882 pounds (400 kg). One report credits a weight "in excess of 1,430 pounds (649 kg)."

David and Melba Caldwell, of the University of Florida, spent many hours watching the behavior of bottlenoses at Marineland. They concluded that courtship is violent, the male and female bumping heads forcefully. Finally, "the two animals swim straight toward one another, head on, until they either hit full force . . . or glance off and slip down each other's

side. Intromission is rapid and takes place under water almost belly to belly. . . ."

The newborn calf is 38.5 to 49.6 inches (98 to 126 cm) long and weighs 20 to 25 pounds (9 to 11 kg). This variation possibly includes some premature calves. Fond of warm, shallow inshore waters, the bottlenose ranges in summer as far north as Cape Hatteras, North Carolina, and in the west to Point Conception, southern California; also year-round off Hawaii. Florida has long been the center of a dolphin live-capture industry.

The bottlenose is a distinctly social species, usually traveling in groups of as many as a dozen, occasionally in aggregations of several hundred. Most populations evidently do not migrate, although they go where they can find food.

In the wild, the bottlenose feeds on squid, shrimp, and a wide variety of fish. In some waters bottlenoses habitually follow shrimp boats, recovering what the shrimpers discard or miss. They often hunt as a team, herding small fish, such as menhaden, ahead of them and picking off the stragglers. One day as author Scheffer was wading along a sandy Mexican beach, three bottlenoses drove a school of foot-long (30-cm) silvery mullet into the shallows around his feet. The dolphins wheeled to make another rush, scraping the bottom as they turned.

A blindfolded bottlenose can tell, from the nature of the echoes, the difference between a copper plate and an aluminum plate of the same size and shape.

Spinner Dolphin *(Stenella longirostris)*

This agile dolphin can leap from the sea, spin two and a half times around the axis of its body, and plunge into the water within a second's time! Thus its name—though no one knows *why* it spins. Spinning is evidently not a courtship display, for mature and immature animals of both sexes engage in it.

Little was known of spinner dolphins until the 1960s, when commercial fishermen in the tropical Pacific Ocean began to kill tens of thousands each year in newly invented nets for catching yellowfin tuna. The dolphins,

trapped along with the fish in the slowly closing, or pursing, nets would suffocate.

The Marine Mammal Protection Act of 1972 nearly eliminated killing by U. S. fishermen. Foreign fleets continue to take large numbers, but new U. S. restrictions on tuna imports may help reduce the kill of dolphins. Originally estimated to number more than 5 million in the eastern Pacific, fewer than 1.5 million spinner dolphins now survive there.

In the eastern tropical Pacific, the maximum known size of

spinner males is 6.4 feet (2 m); of females, 6.1 feet (1.9 m).

The spinner lives in tropical inshore and offshore waters around the world, including Hawaii and the Gulf of Mexico. Little is known of its migration. In the eastern tropical Pacific its home range is roughly circular, about 200 to 300 miles (370 to 555 km) in diameter; it moves seasonally several hundred miles onshore (possibly in fall and winter) and offshore (possibly in spring).

The small, white earbones (otoliths) of fish are hard,

resistant to digestive juices, and characteristic for each species. Thus, an expert who examines earbones from the stomach of a sea mammal can tell what fish were in the animal's last meal. A study of earbones from spinner dolphin stomachs disclosed that lantern fish—tiny, pelagic, tropical fish—composed more than half the dolphins' diet. At times spinners evidently feed more than 800 feet (243 m) below the ocean surface, for the otoliths of deep-sea smelt have been found in their stomachs.

Killer Whale (*Orcinus orca*)

Unlike most cetaceans, this one preys on other warm-blooded animals. The killer whale is the marine counterpart of the tiger and wolf. A man in a boat off the California coast once saw a big male killer whale leap clear of the water, holding a full-grown sea lion crosswise in his jaws; no small feat, when a bull sea lion may weigh 600 pounds (272 kg).

Another man watched a killer chasing a sea lion in the rocky shallows of the California coast. The quarry suddenly changed course, causing the killer whale to collide head on with a large rock. Stunned and quivering, it lay ten minutes before it recovered its senses and swam away.

The killer is the speedster among whales, attaining 29 miles (46 km)

an hour, and is remarkably agile. Skana, a killer whale in Vancouver Public Aquarium, British Columbia, would hurl her powerful body into the air and touch a goal 23 feet (7 m) above the water.

The largest male measured 32 feet long (10 m) and weighed about 9 or 10 tons (8.2 to 9.1 t); the female, 28 feet (9 m) and 5 to 6 tons (4.5 to 5.4 t).

The newborn calf is about 8 feet (2.4 m) long and weighs an estimated 400 pounds (181 kg). The Japanese have reported a 9-foot (2.7-m) fetus and a 9-foot calf.

Worldwide in distribution, the killer whale ranges north and south to polar ice, especially near coasts. In waters of Washington and southern British Columbia, where killer whales have long been studied, the animals cruise in family units or pods of about ten.

All-white (albino) killer whales have been seen with normal ones. Their travels are reminiscent of the seasonal, regular movements of wolf packs in the forest, except that the whales move continuously, day and night.

Japanese whaling captains who examined the stomach contents of 364 killer whales found, in order of occurrence, fish (cod, flatfish, sardines, salmon, tuna, and others); octopus and squid; dolphins (Dall's, blue-white, and finless black); whales (beaked, sei, and pilot); and seals (harbor and ringed).

Dale W. Rice, whale expert for the U. S. Government, reported on the stomach contents of ten killer

whales collected between Kodiak Island, Alaska, and southern California. He found parts of at least three California sea lions, four Steller's sea lions, seven elephant seals, two harbor porpoises, two Dall's porpoises, and one minke whale, as well as fish and squid.

When the recorded voices of killer whales were transmitted underwater near migrating gray whales, the grays appeared to be frightened. They swam away from the sound source. Pure-tone sounds and random noise did not bother them.

Zoological evidence tells us that, although killer whales frequently harass larger whales by slashing at their flippers and flukes, they do not often kill one. There are no confirmed records of attacks by the killer upon human swimmers.

Short-finned Pilot Whale *(Globicephala macrorhynchus)*

This big, social dolphin is named from its habit of following a leader or "pilot," usually the largest male in a group. Whalers used to take advantage of its habit to drive pilot whales into shallow waters where they could easily be killed.

When 28 pilot whales ran aground on San Clemente Island, California, three scientists flew to the scene and examined the bodies. Finding no diseases or poisons, they concluded that the whales had followed the leader aground in hot pursuit of squid—

their preferred food—then spawning in the shallows.

The record size for male pilot whales is 20.2 feet (6 m) and about 3.2 tons (2.9 t); for females, 16.8 feet (5 m), weight unrecorded.

Pilot whales range throughout tropical and cool, but not polar, seas of the world. Two species inhabit North American waters— the long-finned pilot whale (*G. melaena*) in the cool temperate Atlantic and the short-finned in warmer waters of both Atlantic and Pacific. Off the Pacific coast of North America, the pilot whale is

seen as far north as the Gulf of Alaska. In the Atlantic, it ranges to Newfoundland.

Navy scientists trained a male pilot whale named Morgan to dive with a grabbing tool clamped onto his beak. Thus Morgan could recover "lost" torpedoes from the sea bottom. Once he dived to the astounding depth of 1,654 feet (504 m), surfacing in 13 minutes.

When collectors for an aquarium tried to capture a baby pilot whale,

its mother kept pushing the little one away from the capture boat. The baby spouted streams of bubbles from its blowhole. Once landed on deck, it squeaked and chirped continuously, while its mother frantically circled the boat. The men, concluding that the baby was still nursing, restored it unharmed to its mother.

Bimbo, a pilot whale that lived for eight years in Marineland of the Pacific, developed symptoms which, in a person, would be called neurotic. Mercifully he was turned loose at sea.

Harbor Porpoise (*Phocoena phocoena*)

This shy little porpoise, smallest of North American cetaceans, usually is seen at a distance in some secluded harbor as it breaks the surface and quietly blows. *P. phocoena* never approaches a moving ship to sport at the bow wave, and it never leaps clear of the water. Its populations have probably suffered greatly from 20th-century ship traffic, coastal construction projects, and water contamination.

The largest size for both sexes is about 6 feet (1.8 m) and 200 pounds (91 kg). The newborn calf is about 29 inches (74 cm) long and weighs 12 to 16 pounds (5.4 to 7.3 kg). It may live 13 years or longer.

The harbor porpoise ranges from the Arctic Ocean southward along the Pacific coast to southern California and along the Atlantic coast to the Delaware River. It frequents cool coastal bays and the mouths of large rivers. Its migrations are more inshore to offshore than north to south.

The harbor porpoise feeds on a wide variety of small fish and squid, including bottom dwellers. Six porpoises that suffocated in a sea-bass net on the bottom of Morro Bay in California had been feeding in 90 feet (27 m) of water.

Two men in a boat near the California coast saw an attack by several hundred harbor porpoises on a great school of sardines moving just beneath the surface. Repeatedly, five to seven porpoises would form a rank, side by side, to charge madly through the school. One porpoise might devour as many as a dozen fish in a single rush. Several animals circled the school continuously as though they were "riding herd."

The great white shark (*Carcharodon carcharias*) preys upon the slower harbor porpoise. In the stomach of a one-ton (0.9-t) shark caught off eastern Canada were the remains of three porpoises—evidently grabbed from behind, for the tail section of each had been severed by the shark's teeth. A hunter, aiming to shoot a harbor porpoise near the coast of Nova Scotia, was defeated by a 15-foot (4.5-m) great white shark. The shark rushed in and bit the porpoise in two, leaving the frustrated hunter with only the head end.

A zoologist who tested a harbor porpoise wrote that its underwater hearing sensitivity was "among the highest ever measured among animals."

Harbor porpoises are seldom displayed in aquariums. They are easily frightened, even though they may have been gently treated for many months.

Some zoologists classify the harbor porpoise and Dall's porpoise in a family separate from the delphinids, partly because these porpoises have spade-shaped rather than conical teeth.

A closely related species (*P. sinus*) lives in the warm waters of the Gulf of California, Mexico, where it is called *vaquita,* Spanish for "little cow." *P. sinus* is now endangered, mainly because it is taken in nets set for fish.

Dall's Porpoise *(Phocoenoides dalli)*

This handsome, black-and-white porpoise is often seen in North American waters. Although it lives only in the North Pacific Ocean and adjacent seas, it ranges from the Bering Sea to Baja California. Moreover, it often plays at the bow of a moving ship—racing alongside, seeming at times to dare the pilot to run it down.

During the 1980s, Japanese gill-net fishermen in the western North Pacific were accidentally catching and killing an estimated 25,000 Dall's porpoises each year.

This kill increased to as many as 40,000 annually.

The largest male known was 7.2 feet (2.2 m) and 480 pounds (218 kg); females are slightly smaller. The newborn calf is about 39 inches long (99 cm) and 84 pounds (38 kg). The young may be born from June to October.

In the northern part of their range, these porpoises move in spring from the Gulf of Alaska to the Bering Sea and return in the fall. Off southern California, they are most numerous in the fall and winter. They travel in groups of 2 to 20, though mariners have seen

concentrations estimated at 2,000 to 5,000 animals.

The Dall's eats squid and fish such as saury, hake, herring, jack mackerel, and deep-ocean and bottom fish.

Although its swimming speed has not yet been timed, the Dall's porpoise may prove to be the swiftest of all small cetaceans. Compared with related dolphins, it has a larger heart, more blood, and greater capacity for holding oxygen in its blood. And in captivity it eats twice as much food

as does a bottlenose dolphin of the same weight.

Because of its active, excitable nature, the Dall's had, up to 1965, never been captured and held for more than a few days. Then Navy scientists at Point Mugu, California, succeeded in keeping Marty, a 264-pound (120-kg) male, for 21 months.

The Makah Indians of western Washington called the Dall's porpoise "broken tail," in reference to the large muscle mass on its tail stock. The hump conceivably boosts the animal's speed through the water.

Family Monodontidae

These medium-size cetaceans—the narwhals and white whales—dwell only in the Arctic Ocean and adjacent seas. They lack a dorsal fin, have a small, rounded, beakless head, and have a short mouth cleft that resembles a small smile. The seven neck vertebrae are separate, whereas in dolphins and porpoises (family Delphinidae) some vertebrae are fused. As a consequence, the monodontids have a suggestion of a "neck."

Of all the toothed cetaceans, only the two species represented in this family have become adapted to breeding in water that is chilled by melting ice. They have no counterparts in Antarctica.

In their rigorous environment, they have learned, so to speak, the advantages of sociability. All are gregarious.

Narwhal (*Monodon monoceros*)

The narwhal's common name means "corpse whale" (*nahvalr*) in old Norwegian, from the resemblance of its mottled skin to that of a drowned person.

The male narwhal is unique among cetaceans in having a tooth that continues to grow until it becomes a lancelike, hollow, spiral tusk up to 9 feet long (2.7 m) and weighing 18 pounds (8 kg). Male and female have only two teeth (the upper front), and only in the male does the left one normally become a tusk. It is thought to be a secondary sex characteristic like the antlers of deer, useful in territorial jousting. Once a tusk was found jammed inside the broken shaft of another one, as though two males had rammed each other head on.

The largest male measured was 15.4 feet (5 m) long and 3,528 pounds (1,600 kg); the female 13.1 feet (4 m) and 1,984 pounds (900 kg). The newborn calf, born in summer, is 5 to 5.5 feet (1.5 to 1.7 m) in length. Narwhals live only in

the Arctic Ocean and adjacent seas. Russian scientists based on a drifting ice station saw narwhals within five degrees of the North Pole. In the Canadian Arctic, the largest numbers are seen in Lancaster Sound, in Repulse Bay, off northeastern Baffin Island, and off northern Southampton Island. It is the dream of many a Canadian Eskimo to find a *savssat,* or place where a pod of narwhals has gotten trapped in a bay by sea ice. Here these little whales can easily be taken for use as dog food, human food, lamp oil, and articles of handicraft. One traditional use for

the tusk of the narwhal male was as material for the *unang,* a light harpoon used in hunting seals. One tusk was long enough to make the entire shaft of the implement.

Several thousand narwhals may migrate as a group. In winter, they move ahead of sea ice as far south as Hudson Bay and Labrador. Narwhals in Canada and northwestern Greenland number about 30,000. They are not common in northern Alaska.

They feed on deep-living bottom crustaceans and squid and also take the abundant polar cod and Greenland halibut. Like the other odontocetes, narwhals crush fish between their powerful jaws and swallow without chewing.

In September 1969, the New York Aquarium became the first to exhibit a live narwhal, a calf that lived only a few weeks.

White Whale *(Delphinapterus leucas)*

This, the only all-white cetacean, has been valued for centuries by northern people for its meat, oil, and leather. The newborn calf is brown, gradually paling through shades of gray until, at six or seven years, it turns white.

Docile in captivity, the white whale, or beluga, quickly endears itself. Stefani I. Hewlett, a staff biologist at the Vancouver Public Aquarium, observed the exciting birth of the first white whale to be born in captivity and survive for more than a few minutes. Little

Tuaq—Eskimo for "the only one"—arrived headfirst after mother's labor of 3¾ hours: "Brown, with limp little flukes, it swam vigorously and unaided to the surface, popping straight out of the water to past its pectoral fins at least three times. 'Breathe, please, breathe,' prayed Stefani. The baby whale took its first breath on July 13, 1977." It lived 16 weeks.

An adult of record length, 16.4 feet (5 m), weighs 3,528 pounds (1,600 kg). Females are slightly smaller than males. Seventeen newborn calves from Baffin Island had an average length of 5.2 feet

(1.6 m) and weight of 172.6 pounds (78 kg).

Females begin to breed at five years, males at eight years. After mating in May, gestation lasts about 14½ months. Calving often takes place in the warm waters of river estuaries. The calf nurses for a long time—about two years. Growth layers in teeth and jawbones indicate a maximum life span of 30 years.

The beluga ranges as far south as Bristol Bay, Alaska, and the Gulf of St. Lawrence, Canada. It often ascends rivers and has been seen in

the Yukon River 600 miles (965 km) from salt water. Whites often travel in groups of hundreds. Near Canada's Somerset Island, more than 1,000 individuals have massed at calving time. White whales talk among themselves. Their voices carry above water and are heard by hunters in boats.

At a turn in the St. Lawrence River 60 miles (96.5 km) downstream from Quebec, fishermen have caught as many as 320 white whales in a day in the world's only whale trap. It is a seasonal, corral-type structure built of slender poles

Family Physeteridae

A massive forehead and a short, narrow lower jaw give the sperm whales a "chinless" appearance. The spermaceti organ, or case, a unique reservoir of transparent, waxy oil, fills most of the forepart of the head.

The family contains three species—sperm whale, pygmy sperm whale, and dwarf sperm whale—all of which may be seen along the North American coast. The pygmy and dwarf forms (which some authorities place in a separate family, Kogiidae) are not known to exceed 11 and 9 feet (3.4 m and 2.7 m) respectively. Record length for the sperm whale is 66 feet (20 m).

Sperm whales roam the deep waters of all the oceans, though they seldom approach the polar ice fields. Off North America in summer, males are sighted as far north as the Bering Sea and Davis Strait, females and immatures as far north as latitude 50°.

In autumn, both groups migrate southward. Migration routes and areas of concentration are shown on the map.

jammed into the river bottom. The North American population of whites in recent years was at least 30,000. Of this number 10,000 were regular visitors to Hudson Bay, another 10,000 to Lancaster Sound, Canada, and 10,000 to Alaskan waters.

White whales feed near bottom in shallow waters, where they take salmon, capelin, cisco, pike, char, cod, and other fish, and squid, crustaceans, and sandworms.

Zoologists of the Alaska Fish and Game Department learned that hundreds of white whales habitually gather in the Kvichak River around the first of June each year to gorge on valuable young salmon. So the zoologists placed in the river a powerful sound transmitter that played back the tape-recorded voices of killer whales. When the white whales first heard the sound, they "turned immediately . . . and swam directly out of the river against the strong incoming tide." This system of biocontrol—the control of life with life—quickly became standard procedure.

Sperm Whale (*Physeter catodon*)

When Herman Melville created *Moby Dick,* the story of a bull sperm whale, he could not have chosen a more exciting hero. Fiercely aggressive, outweighing the female more than two to one, the bull sperm earned the fear and respect of small-boat whalers of the 19th century. They learned to their sorrow that he might suddenly turn and attack the ship that brought his tormentors.

The sperm whale may be the most numerous of the great whales, though it was hunted intensively until about 1980.

Original worldwide numbers may have been as high as 3 million; current estimates range from under 100,000 to more than a million.

The enormous, boxlike head of the sperm whale sets it apart from all other species. The head may contain three or four tons of spermaceti, a substance valued as a lubricant for fine machinery. The blowhole slit is skewed, not positioned at a right angle to the backbone as in beaked whales and

dolphins. Moreover, it is on the left side of the head. Small flippers, situated near the rear of the head, could casually be mistaken for external ears. There are 18 to 28 functional teeth on each side of the lower jaws. The upper teeth are few, weak, and nonfunctional. Most never erupt through the gums.

A male may reach 66 feet (20 m) and 63 tons (57.2 t); the female, 41 feet (12 m) and 26 tons (23.6 t). There have been questionable reports of males up to 69 feet (21 m). The newborn calf measures about 13 feet (4 m) long and weighs about a ton (0.9 t).

Sperm whales are polygamous, and individuals group themselves roughly by age and sex. A report of the National Marine Fisheries Service says sperm whales are found "singly or in groups of up to 35 or 40 individuals. Older males are usually solitary except during the breeding season." During the rest of the year large groups may include bachelor bulls (sexually inactive males) or may be "nursery schools"

containing mature females and juveniles of both sexes.

The female matures sexually at 8 to 11 years; the male at about 19 years. The maximum known age is 77 years. The gestation period is 14 to 15 months, and the calf nurses for about 2 years.

The sperm whale feeds mainly on squid and on octopus and deepwater fish. Workers at the whaling station at Faial Island, Azores, found in the stomach of a sperm whale the undigested remains of a giant squid weighing 405 pounds (184 kg) and 34.4 feet (10 m) long. Zoologist Robert Clarke, who examined the squid, wrote: "It seems that a whale capable of taking the squid . . . would have little difficulty in swallowing a man, unless it be that a man might slip less easily down the gullet than a mucous-smooth squid."

Off California, wrote whale expert Dale W. Rice, these whales often take roughscale rattails, sablefish, brown cat sharks, longnose skates, lingcod, Pacific hake, rockfish, and king-of-the-salmon (a ribbonfish rather than a true salmon).

By listening through directional hydrophones to the echolocation clicks made by a sperm whale, zoologists tracked the animal to a depth of 8,200 feet (2,500 m). "It seems likely that the depth to which a sperm whale can dive is limited only by the length of time it takes to get down and back," wrote Rice. Two large bulls were shot off Durban, South Africa, in water nearly two miles deep. They had been down 80 minutes and had fresh, bottom-dwelling sharks in their stomachs.

Scientists offer two main theories for the purpose of the spermaceti organ. It may focus and reflect sound. And it may serve as a cooling organ, thus diminishing the whale's volume and its buoyancy during prolonged dives. If the latter theory is true, the animal starts to "blush" as it dives, simultaneously sucking cold water through its right nostril. Cooling the spermaceti only 5.4°F (3°C) would, in theory, make the body neutrally buoyant.

Family Ziphiidae

These are the beaked whales, diverse in form and habit, exploiting many food niches from the tropics to the polar ice edges. Moderate in size, they have a narrow, tapering snout which, in some of the species, resembles the neck of a bottle. The rear edges of the tail flukes have shallow notches, a distinguishing feature. Beneath the throat are two distinct grooves, which almost meet in front and diverge as they continue onto the chest.

The dorsal fin is small and positioned near the rear, while the flippers, also small, are positioned low on the underbody. Most beaked whales have only one or two pairs of functional teeth. In the female these are typically buried in the gums. An exception occurs in the rare Shepherd's beaked whale (*Tasmacetus shepherdi*) of the South Pacific, with about 48 pairs.

Whitish, crisscross scratches and scars are commonly seen on the skin of beaked whales of both sexes. Some are tooth marks caused by in-group fighting, while others are marks inflicted by squid beaks.

Beaked whales hunt squid and deep-sea fish, which they crush between the lower jaw and the rough, horny palate, the roof of the mouth.

Baird's Beaked Whale (*Berardius bairdii*)

Before shore whaling ended in 1971 on the Pacific Coast of North America, whale hunters would occasionally shoot a Baird's beaked whale as a consolation prize, for this slender whale was considered barely large enough to pay expenses. Canadian whalers at Coal Harbour, British Columbia, used to kill one or two a year; Japanese whalers are still killing about 60 of them a year.

The longest male measured was 39 feet (12 m) and weighed some 13 tons (11.8 t); the female is larger, reaching a record 42 feet (13 m) and about 14 tons (12.7 t). Length of the newborn calf is

about 15 feet (5 m), its weight a little more than a ton.

These whales inhabit the North Pacific region from the Bering Sea southward to Japan and southern California. They travel in tight schools of up to 30 individuals and are wary of approaching ships. Surely they migrate, though the migration pattern is poorly known. Their food consists of deepwater fish, squid, and octopus. Mating is believed to take place in the spring. The calf is born in midwinter, after a gestation period of 17 months.

OVERLEAF: *Trained dolphins dance an aerial ballet in Hawaii's Sea Life Park. In the wild, schools of these exuberant mammals leap and splash for sheer joy—and possibly to signal other dolphins.*

Meat Eaters

ORDER Carnivora

WOLVES AND WILDCATS, bears and weasels. These and many other meat eaters compose one of the most fascinating orders of mammals, the carnivores. They are the hunting beasts of the animal world. Consisting of some of humankind's closest companions—as well as our fiercest competitors—this order spans a roster of species that have held our attention for centuries.

We can debate forever whether our ancestors' first attitude toward carnivores resulted from fear of becoming their prey, from hatred for them as competitors for food, or from appreciation of them as bearers of warm and lustrous furs. Toward a single carnivore, the wolf, one can still see all these attitudes present in humans today—along with a strong social bonding to the wolf's direct domestic descendant, the dog. Myths, legends, fairy tales, and horror stories involving wolves permeate most of the cultures in the Northern Hemisphere, the wolf's original range. And in varying degrees it has been the same with the bears, cougars, weasels, wolverines, and other members of the order.

Not many carnivores can be considered placid. Their need to feed on flesh compels them to dig, scurry, ambush, pounce, bound, lope, climb, or leap. When aroused, they may huff, scream, bark, screech, or howl. Generally they are sleek and strong, quick and alert, their senses finely tuned. Seldom is a carnivore accused of being dull.

Carnivores are a varied lot, with five major groups, or families, present in North America: the dogs (canids), bears (ursids), raccoons (procyonids), weasels (mustelids), and cats (felids). And they have been a long time in the making. They separated from the ancestors of their prey, the plant eaters, about a hundred million years ago and have since developed into the animals we know today. Over time

they evolved their own specializations—massive claws in the bears, long-distance communication in wolves, and extreme dexterity in raccoons.

The most distinctive traits common to the members of this order are those associated with their mostly carnivorous diet. Their powerful jaws are specialized for biting and tearing. Their highly differentiated teeth allow efficient butchering and consuming of prey. In most carnivores, the last upper premolar and first lower molar are developed as carnassials, meat-cutting teeth that scissor past each other to slice tough sinew. Massive molars crush bone, and large pointed canines—fangs—grasp and tear. David Mech, one of the authors of this chapter, once watched a wolf put its fangs to spectacular use in helping its pack capture a large moose. The wolf grabbed the moose by its rubbery nose and held on. The moose raised its head and shook it from side to side, swinging the wolf back and forth in the air. Meanwhile, the other members of the wolf pack kept tearing at the rump of the moose until they pulled their quarry down and finished it off.

After killing its prey, a carnivore is well equipped to make the most of it. In a matter of minutes a large predator tears into its catch and gorges on it. Compared with plant material, flesh is highly digestible. Thus the carnivore's digestive system has a relatively well-developed foregut—useful for food storage, digestion, and absorption—and a short hindgut. Its stomach is large and simple. A wolf, for example, can eat almost 20 pounds (9 kg) at a time, and the food is assimilated within a few hours. Several engorgements during the first day after the kill, then, can allow the carnivore to quickly take advantage of what usually is a hard-earned capture. Although carnivores can feast when they make a kill, they often must contend with famine between kills. Several environmental factors affect the number, availability, and vulnerability of prey, and these factors often work against the predator.

After feeding, a carnivore may have to travel many miles and survive many days before locating another animal it can capture. During that time it may lose a great deal of weight. Therefore, the carnivore must try to profit as much as it can from its temporary

scavenging on the carcass. After hollowing out a cavity in the abdomen of the moose, the fisher took up residence in the hollow. As the carnivore alternated between sleeping and eating, its home grew larger and its larder dwindled. Eventually it ate itself out of house and home.

Just as a carnivore's digestive system and feeding habits are adapted to efficiently process a diet of flesh, so too are its feet adapted to efficiently run down prey animals. Carnivore feet are strong and supportive—even in fluffy snow. Whereas many plant-eating prey animals have sharp, heavy hoofs for defense against and escape from carnivores, the meat eaters have kept the generalized use of most of their toes. This enables their feet to retain flexibility over almost any terrain—muddy areas, fallen timber, or rocky ridges—but still fosters fleet-footedness.

The front feet of most carnivores are reasonably well adapted for food handling. Even in canids, which probably have the least flexible front feet, the paws are useful for holding down bones and small prey while the animal strips or tears chunks of meat from them. The front feet of cats have sharp claws that spring out and puncture prey. Bears possess massive, pointed claws both for hauling down their prey and for digging insects and rodents out of the ground. The front feet of raccoons and weasels are highly flexible, allowing these animals to readily manipulate their prey while feeding.

These physical and behavioral adaptations to a flesh-eating economy would be of little value if carnivores did not have well-adapted senses for finding their prey. Some of the feats of the carnivore sensory system are spectacular. Wolves, for instance, can hear other wolves howling a distance of at least six miles (9.5 km) away in forested terrain. Most carnivores can tell in which direction a prey animal has run merely by scenting its trail, which may be several hours old. This means the pursuer can detect minute changes in the scent gradient along the trail, a gradient whose changes result from differences of only

A wolverine's dental arsenal typifies that of most carnivores. The upper jaw carries (from left in skull diagram) incisors, a canine, premolars—including the large carnassial—and one molar.

bonanza. Various groups solve the problem in different ways. Members of the dog family bury excess food beneath the ground or snow for later retrieval. Cats and bears protect their kills by scraping leaves and ground detritus over them and returning frequently until the food is eaten. Mech once saw a huge grizzly bear lying completely across a bull moose on which the bear had been feeding. It was hard to imagine anything on earth trying to usurp that animal's food supply.

Weasel family members may hide right with their prey in some cavity or burrow and stay there until all the meat is gone, sometimes for as long as five or six days. The record tells of a fisher—a large member of the weasel family—that found a frozen moose and began a few seconds in the life of the scent. Some dogs can distinguish identical twin human beings by odor and can scent a human fingerprint two to three weeks old.

Most carnivores can see well in the dark because their eyes contain a well-developed tapetum, or reflecting layer, that allows them to utilize light that would otherwise be lost to them. This ability allows cats, for example, to see at a level of illumination only one-sixth of that required by humans.

These superb sensory abilities, the well-developed behavioral traits, and the highly specialized physical adaptations make the carnivore an efficient meat-processing animal. But to survive over the millennia, carnivores also had to learn to heed certain economic rules. It would not pay a mountain lion, for example, to spend several hours hunting a deer mouse. Nor would a weasel be wise to chase a deer.

Instead, each species of carnivore fits into a certain prey niche, taking primarily prey that are of the size, condition, age, density, or living habits that the carnivore is capable of hunting efficiently.

Furthermore, each carnivore species must be adaptable enough to try new strategies whenever necessary or to suspend hunting if circumstances warrant. One wolf pack in Minnesota, for instance, slept four times as much during a winter in which deer were scarce as when they were plentiful.

Not all flesh eaters are classed as Carnivora. This closely related group of species has counterparts in other mammalian orders: leopard seals of the Pinnipedia, killer whales of the Cetacea. But as a group, members of this order rank among the most typically carnivorous. Hence the name.

Nor are all the carnivores strictly flesh eaters. Most of them are highly evolved, intelligent opportunists, making the most of whatever food is abundantly available. In addition to mammal flesh—and that of birds, invertebrates, fish, amphibians, and reptiles— many of the carnivore species will readily eat many kinds of plants, fruits, nuts, bulbs, or tubers, as well as algae, worms, grubs, and insects. And sometimes carnivores have developed very complicated ways of obtaining their food.

The intricacies of wolf pack cooperation in hunting are legendary. The sea otter actually uses a rock as a tool to open shells of mollusks. Susie, a captive otter used as a subject in Karl Kenyon's long-term study of the species, was given a flat stone along with some unopened clams. A smashing success, Susie grew so fond of the implement she did not let it out of her sight. Later her caretakers found out Susie was also using her pet rock to make more "rocks"—pieces of concrete pounded from the walls of her pool.

Polar bears, the most carnivorous of the North American ursids, are large animals that look cumbersome. But they are both agile and subtle when stalking seals. These bears can pull their ponderous bodies up the steep walls of ice floes or glide silently toward their prey, using irregularities of the ice as cover until they are close enough to strike a blow. When approaching by sea, they can as silently and furtively lower themselves into the water and swim with only a black nose-tip and eyes rippling the surface. Less known is their ability to dive in shallow seas from ice shelves and retrieve kelp from the seafloor. They pick through and eat the kelp.

As a group, carnivores are not only intelligent and able individuals, but they also pass the accumulated "knowledge" of foraging and predation techniques from one to another, especially within family groups or packs. They have, in effect, a highly evolved "cultural inheritance" of great value for exploiting their food resources or surviving in rigorous environments.

The mother bear, for instance, during her long association with her cubs, passes along much information on foraging, the making of dens, and appropriate bear behavior. She is a stern disciplinarian. If a cub does not pay attention, disobeys, or otherwise misbehaves, she may reinforce the lesson with an instructive swat on the rump.

A solitary hunter seizes a morsel—a ground squirrel dug from a burrow. Though grizzlies are able to kill large prey, they mainly subsist on vegetation and carrion as well as small rodents.

Social hunters close in for a kill. As the moose tires from the chase, wolves will attack its rump and sides until it falls. The pack will then gorge on the carcass, leaving only bones, hair, and hoofs.

By the time an individual of the younger generation becomes independent, therefore, it not only has its instinctive abilities but also has learned information from its own experiences and information from its mother and from her mother before.

Carnivore populations in any given area are made up of fewer individuals than populations of their plant-eating prey. Meat eaters must capture and kill what they eat, and when that is consumed, capture and kill again. Obviously they cannot live as densely as their prey species, for they would soon deplete their food resource and perish. And, because of the way carnivores make a living, they must utilize larger areas than plant eaters do. They must also be able to travel relatively great distances in search of food.

To maintain a population density compatible with food resources, carnivores have evolved social systems that help control their numbers. These systems usually involve some form of territorialism, or spacing out of

A quick red fox pounces—perhaps on a mouse or gopher. Scent and sound guide the fox to its target. The fox leaps—and pinions the victim with its paws. Coyotes also employ the tactic.

like animals in a given area. Among wolves and other gregarious carnivores, a pack, or a group of individuals, controls a territory and excludes other packs. Among the solitary carnivores, such as the mountain lion, individuals maintain territories. These territories function as a "home area" or "home range," and each individual or pack usually confines itself to its own territory. This spacing limits the overall population of the species and thus averts strife and starvation.

Territories are maintained in different ways. Wolves howl to announce their presence to neighboring packs, and they urinate to leave scent markers at the boundaries of their area. They also aggressively defend their territorial boundaries against invaders—sometimes killing intruding wolves.

Cats use more subtle means of maintaining territories. It is less desirable for them to fight their own kind. A solitary predator must stay in good physical condition in order to survive; badly injured in a fight, it could not fend for itself and would die. Therefore, a more peaceable system for establishing territory has evolved in the cat kingdom. This involves the marking of home ranges with "scrape stations"—piles of earth, pine needles, or similar material scraped together into visible mounds. After scraping up the mound, the cougar may urinate or defecate on it to leave an olfactory as well as a visual sign that the area is occupied. Another cougar, in search of a home, recognizes the sign and moves on. By avoiding each other, the cougars avoid conflict.

Unlike the canids, most cats do not vocalize to maintain their territories. An exception in North America is the jaguar. Because it lives in dense vegetation where sight and scent signs would not work so well, the jaguar cries out to let others know its whereabouts.

These systems among wolves and some cats probably represent a pinnacle in the evolution of carnivore social behavior. Other carnivores have evolved behavior modified to fit their particular needs and lifestyles. Bears, for example, eat a wide variety of both animal and vegetable matter. Because the diet of these omnivorous carnivores is not restricted to one type of food, their systems of spacing are less rigid. They often range widely, but if food is plentiful, several bears may remain in a small area for a number of days. When bears congregate, as when brown bears gather to feast on salmon, dominance hierarchies develop to control spacing.

Scent glands are well developed in practically all species of carnivores, and all use scent marking to some degree to advertise their presence. It also seems likely that there is a sexual difference in the scent; thus an individual can recognize whether the scent was deposited by a male or a female.

Size of territories differs widely among carnivore species. The least weasel may make use of only a small corner of a woodlot or an unworked area on a farm, while a wolf pack may roam several hundred square miles of northern forest. A bobcat may confine itself to 12 to 15 square miles (31 to 38 sq km), while a male mountain lion maintains a year-round territory of 100 square miles (260 sq km). In general, large carnivores depend on large prey animals and cover big areas; small carnivores hunt smaller prey and prowl smaller areas.

There are exceptions. Wolverines are not large—a big one may weigh only 50 pounds (22 kg). But they roam vast areas. One of the reasons for this is their scavenging way of life. To find enough carrion and other food, the wolverine often must travel far and wide. Other species, including the fisher and the lynx, often travel extremely long distances to find their food. Ordinarily these species stay within a definite area. But if they find food becoming scarce, they must move. Wanderings of many miles each day are not uncommon for either the fisher or the lynx.

Reproduction in the carnivores takes varied forms. Mating behavior differs markedly between the gregarious wolves, for example, and the solitary weasels. A dominance order exists in each wolf pack, and normally the dominant male and dominant female mate. When a female weasel comes into estrus, she probably will breed with the first male she meets. Wolf partners remain together as pack members after mating, and the male helps care for the young. Weasel partners part, possibly never to meet again. The female rears the young alone—the usual pattern for most solitary carnivores.

Delayed implantation occurs in bears and most mustelids; that is, the fertilized egg does not implant and grow immediately. A delay of some months is advantageous because young are born at a time when food is more abundant or, for bears, while the mother is wintering in her den. Nearly all carnivores are born quite small and helpless and are cared for solicitously by their mothers.

The period of juvenile dependency for the larger species is long—nearly two years for cougars. But for most mustelids dependency lasts only three to eight months. Most carnivorous species reach sexual maturity in one to two years. Not every sexually mature

individual necessarily breeds. Again social behavior plays a role. If more than one female wolf in a pack should breed, the pack likely could not support all the pups; a single litter has a better chance of survival. When the dominant, or alpha, female dies, the beta female moves up in the hierarchy and subsequently breeds. In this manner the pack can be maintained at an optimum number.

A young female cougar will ordinarily not breed until she is established on a territory. She can better provide food for her offspring when she is on familiar ground and free from the uncertainties of a transient's life. Once more, social structure operates to avoid overpopulation and its dire effects.

Bears, on the other hand, at times control their numbers by killing each other. As population stress builds up, adult males may kill and eat the young—if they can catch them. And pregnant females lose their fertilized eggs if they do not have a certain amount of body fat by denning time. Both of these conditions directly limit bear populations.

Because carnivores kill to live, their effect on populations of prey species has long been debated—often with more heat than objectivity. One of the authors, Maurice Hornocker, has shown, through a long-term study of cougars and their prey, that the cats normally do not devastate herds of deer and elk but help to keep them within the limits of their food resources. Grazing and browsing animals tend to increase in numbers to the point of eating themselves out of food; catastrophic die-offs result. It takes many years to restore the vegetation and, in turn, the animal herds. Predation by cougars tends to lessen the frequency of violent fluctuations in the populations of prey species.

Members of the order Carnivora vary widely in habits and abilities. We must continue to study them and learn from them. We know they are an adaptable, successful group of species. They merit all those centuries of our attention.

MAURICE HORNOCKER
CHARLES JONKEL
L. DAVID MECH

Maternal duty and pleasure attain a harmony—an Alaskan brown bear nurses her yearling cubs. Rarely can she take her responsibilities lying down, for she is provider, protector, and tutor-by-example.

Family Canidae

Probably people are more familiar with the Canidae than with any other carnivore family because "man's best friend" is a prominent member of that family. But it cannot be said that the dog is a typical member, because of the great variation among breeds of dogs. Neither the Pomeranian nor Great Dane is a typical canid. The German Shepherd comes closer—most canids have its general body conformation: long legs, narrow snout, large, pointed ears, and long tail. North America's wild canids include foxes, coyotes, and wolves (such as the gray wolf opposite).

Canid feet are especially distinctive. Each has four toes that touch the ground along with a prominent rear pad. This pad is not really a heel, for canids walk on their toes, with their heels high off the ground. This makes for a blocky foot that gives excellent support and allows canids to run swiftly. In larger members of the family, the front feet anchor chunks of prey while the animal strips pieces of flesh and gristle from them. The front feet of all canids also are well adapted for digging soil for dens and for food caches. Claws are thick and hefty, not needle-sharp like those of cats, and are of little use in seizing and holding the prey. The claws function mostly for support of the foot while the animal is running.

As might be expected of animals with feet so well adapted to travel, canids are among the most peripatetic members of the carnivore order. Red foxes have been known to travel a straight-line distance of 245 miles (394 km). Wolves may trek 45 miles (72 km) a day.

Long travel is necessary for canids to find enough prey animals that they can catch. Although most canids will eat a large variety of animal life, and plant matter too, they generally rely on mammals they can run down and catch. They may make many attempts for each successful chase, and they thus must cover considerable ground.

Usually the area hunted by each canid social unit is an exclusive territory. For foxes, a territory may cover a square mile or so; for coyotes, several square miles; and for gray wolf packs, 50 to 5,000 square miles (130 to 12,950 sq km). Canids mark these territories with urine and feces, making olfactory "fences" that keep intruders out. Their voices seem to serve as a supplementary means of maintaining territories. Wolves and coyotes howl, bark, and yap. Foxes yip.

Canid territories are occupied by breeding pairs and their offspring. At least for a given year, pairs are monogamous, a trait rare among mammals, nonexistent in other New World carnivores. Leading to this close monogamous relationship is a prolonged courtship and pair bonding in which marking with scent seems to play an important role.

Copulation in canids involves another family peculiarity, the copulatory tie. After the preliminaries, the two partners remain tightly locked together for as long as half an hour. The copulatory tie almost seems to symbolize the relationship between the mated pair, for the two remain close partners and both will care for the pups.

A burrow, rock cave, hollow log, or similar location serves as a maternity lair. Canid pairs produce one litter a year, with each litter averaging four to six young—but sometimes more than ten. The pups' eyes open at about two weeks of age, and they begin eating meat at three to four weeks. They are weaned at five to eight weeks. By fall, they are almost full grown.

Fox pups generally disperse to stake out their territories in early fall, coyotes in late fall, and wolves when one to three years old. This means that wolf packs usually contain several members throughout the year, as pups from consecutive litters remain with the breeding pair. Fox and coyote social units generally contain only the breeding pair after the young depart.

Coyote *(Canis latrans)*

"The coyote is a living, breathing allegory of want . . . always poor, out of luck, and friendless. . . . Even the fleas would desert him for a velocipede." This assessment by Mark Twain, written mostly tongue in cheek, nevertheless does contain some element of truth. Few Westerners have been friends of the coyote. But even its enemies concede its durability. It thrives in the face of all attempts to trap,

poison, or blast it to oblivion. This denizen of the Great Plains is expanding its range eastward to the Atlantic, partly because of extirpation of the wolf. New Englanders call it the "coydog" or brush wolf. But it is the same old coyote of lore and legend.

The name "coyote" comes from the Aztec *coyotl.* Its Latin name means "barking dog." Adults, 2 feet (61 cm) high at the shoulder, are 3.4 to 4.3 feet long (105 to 132 cm) and weigh 20 to 50 pounds (9 to 23 kg).

Tough and wiry, with keen senses and a quick wit, the coyote

adapts readily to almost any habitat. It has steadily extended its range as it moved off its native prairies in the late 19th century, through the Great Lakes states and into Canada, then New England, New York, and as far south as West Virginia. By the 1960s *C. latrans* had reached Louisiana, Mississippi, and Georgia. Now it is in the Carolinas and Florida. The coyote's range blankets the central parts of the continent south of Canada; it reaches from Alaska south as far as Panama, and is starting to push

into South America. And it is fast—up to 30 miles (48 km) an hour in a dead run. Coyotes hunt alone (like the mouse stalker above) or team with others to grab a meal. They will eat anything—from rabbits, rodents, and carrion (most of their diet) to watermelons and insects.

Coyotes are monogamous and prolific. The female bears five or six pups each spring; both parents share in their upbringing.

Ten-week-old coyote pups relax at the mouth of their den. Such dens, often remodeled badger holes, may tunnel 30 feet (9 m) into a hillside.

Calls of the Wild

A Montana coyote fills a frozen morning with its haunting howl. Others howl a response; two or three can sound like a dozen. The wolf shares the howling habit—and so, sometimes, does their housebound cousin the dog.

Why howl? Theories abound, but they share a common premise: To socialize, one must communicate, and howling is communication. Wolves may howl together at dusk to inspire one another, as if with martial music, for the hunt ahead. Perhaps they howl to rally stray members of the pack. Or howling may be an audible "fence" around a pack's territory; hearing it, other packs stay clear and reply with a fence of their own.

Using sirens, recordings—even human howls—researchers coax responses from wild canids. Here, taped in the wild and transcribed as musical phrases, are typical howls. They vary as human voices do—from a coyote's brief soprano to a gray wolf's alto. The red wolf's grace notes and inflections produce the most varied melody.

Canids communicate in myriad ways. Some mark their territory with scent posts; a squirt of urine on a tree or rock can tell others the age and sex of each animal that "signs in." Snarls, barks, and growls enrich the vocabulary during closer encounters. A complex body language conveys many meanings.

One wolf flushes a goose, which flees straight into the jaws of a second wolf. Do they communicate on such stratagems? Most experts say no. As with talkative humans, success often takes luck.

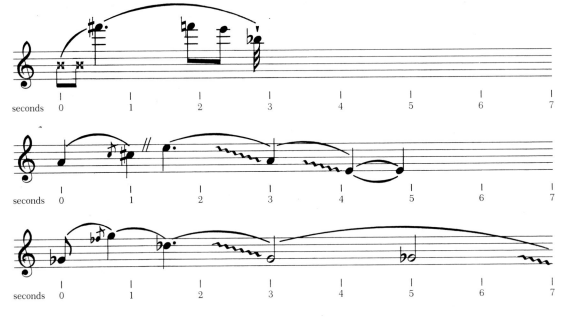

Coyote's howl opens with staccato barks, goes right to its highest note, and ends in a few seconds.

Red wolf's howl builds slowly, hits a melodic high, then trails off evenly.

Gray wolf has the longest howl, a mournful slide from an early high down through an octave or more.

The wolf's lifestyle requires communication—and thereby hangs a tail. This expressive appendage can send a wide range of messages to other wolves.

Held aloft, the tail asserts its owner's dominance. A tail hanging loosely indicates a wolf relaxed and under no social stress. Slight upward curve may be the beginning of a full-fledged threat.

Tail tucked under signals submission. Wags, twitches, even changes in fullness add subtleties to the rich vocabulary of the tail.

Gray Wolf (*Canis lupus*)

At home in almost any habitat, the gray wolf once numbered 20-odd subspecies throughout the continent. Centuries of trapping, gunning, and poisoning—usually for a bounty—have erased several races and backed the rest into an ecological corner. They have remained common in parts of Canada, Alaska, and Minnesota; small populations have returned to Montana, Michigan, and Wisconsin. In 1995 reintroduction efforts began in Idaho and in Yellowstone National Park.

Lone wolves are rare; most hunt in packs of a half dozen or more.

Gray wolves can lope after a deer, moose, or caribou for miles until the exhausted quarry turns to make a stand. Slashing hoofs are often more than a match for the wolves. Only one chase in ten may succeed.

When feeding begins, so do the food fights—mostly bluff, for each wolf knows its place in the pack's hierarchy. At the top reigns the alpha male; to him go the choicest parts. To him also go the duties of leadership in the hunt or in confrontation with intruders. And to him goes the right to mate.

Excitement ripples through the pack at whelping time. All may

Present Range
Former Range

help to rear the one to eleven pups, tending them, fetching food, even adopting the orphans if the parents die. An adult may return from a kill with a full gut. Pups nipping at the corners of the adult's mouth may stimulate the older wolf to disgorge a meal for the pups.

Fully grown gray wolves measure 4.5 to 6.5 feet (137 to 198 cm), tails included, and usually weigh 75 to 110 pounds (34 to 50 kg). The fuzzy pups reach full size in about a year. Then coarse guard hairs cover an underfur so dense that the animal can sleep on snow at minus 40°F (-40°C).

Gray wolves in Minnesota tussle in a display of dominance (left). Yet the winner's hang-dog tail indicates that this is just a temporary put-down.

Flattened ears, bared teeth, raised hackles, and throaty growls keep the wolf pack's hierarchy clear as the animals feed on a deer (below). The dominant male claims first feeding rights, gulping down perhaps a fifth of his own weight in a day. At small kills, others in the pack may go hungry, but he stays healthy to lead future hunts and thus ensure food for all in the long run. And even a hungry alpha may fight off other adults only to defer to a small pup.

Red Wolf (*Canis rufus*)

It was the wolf that no one knew. Its home was not the tundra or north woods but the gentle hills and bottomlands of the South. Like a ghost of things long past, it lurked on the edges of our farms and cities through the middle of the 20th century. Then, despite an eleventh-hour surge of sympathy and research, its last wild populations succumbed to a unique process of genetic erosion.

Old specimens, locked away in museums, show that this species once occurred as far north as Pennsylvania and as far west as central Texas. John James Audubon coined its name after seeing some of that color near Austin. In most of its range, it resembled *C. lupus* in color but was smaller: 40 to 75 pounds (18 to 34 kg), with a narrower physique and shorter fur. Its prey, mostly rabbits and rodents, was smaller than that of the gray wolf. So was its home range, about 18 square miles (47 sq km). Pairs established territories, mated in winter, and produced four or five young in the spring.

Human persecution caused a steady contraction of the red wolf's range. Meanwhile, the more prolific coyote pushed in from the west and north, its way opened by human environmental disruption and by elimination of the larger red wolf. As coyote populations expanded,

they interbred with and eventually absorbed the scattered, remnant pockets of red wolves.

By 1970 the only pure red wolf population was found along the Gulf Coast in southeastern Texas and southwestern Louisiana. Efforts to save this population failed; the U. S. Fish and Wildlife Service began a Dunkirk-type evacuation. Of those saved, 14 survived to become the founders of a captive population. That group,

☐ Original Range ☐ In 1950
☐ In 1930 ☐ In 1970

in turn, was the source of a reintroduction project in coastal North Carolina, starting in 1987. More than 50 wolves now hunt and breed there, much like their prehistoric ancestors.

Arctic Fox (*Alopex lagopus*)

In a lineup of North American foxes, this one would be singled out because of visible adaptations to an Arctic climate. Compact body, short legs and muzzle, and small, rounded ears—forms that conserve body heat—mark this polar species.

As winter begins, the arctic fox sheds its brown coat for a thick white one. A minority wears bluish gray fur in summer, lighter bluish gray in winter. Day length—not temperature—triggers molt. Active all winter, arctic foxes hole up in a snowbank or den only during severe storms. They feed mostly on lemmings. Probably the widest ranging terrestrial mammals, they will travel more than 1,000 miles (1,610 km) in search of food. In winter they scavenge polar bear kills on the pack ice. Near the sea they get a seasonal bonanza of fish, seabirds, and their eggs.

Close teamwork helps this fox cope with a precarious, boom-bust environment. Pairs form in late winter and stay together through the vixen's 51- to 57-day gestation. They den under a gravelly mound.

The best den sites are reused and enlarged year after year until the entire mound is honeycombed with tunnels and portals. On top, the fox keeps a watch post, a place from which to spy out danger or prey.

Large litters, averaging six or more pups, help the species recover from famines brought on by periodic lemming population crashes. Feeding their litter may keep both parents working 19 hours a day.

Red Fox (*Vulpes vulpes*)

Unlike most carnivores, the red fox has expanded in numbers and range since European settlers arrived. It is now the most abundant and widespread fox in North America.

Before settlement, only gray foxes were common in the dense eastern forests. Clearings for farms provided the habitat that red foxes prefer—open spaces bounded by protective stands of trees and brush. The opening of fields also increased the number of voles, mice, and rabbits—all red fox prey.

European settlers also brought a love for fox hunting. Stymied by the gray fox's tree-climbing tactics, they imported European red foxes. Only later was it discovered that red foxes already lived here. Most mammalogists now classify European and American red foxes as one species. Though most bear the golden red color that inspired their common name, red foxes also come in brown, black, and silver—at times in the same litter. A bushy white-tipped tail accounts for a third or more of the fox's average 42-inch (107-cm) length.

Opportunistic and omnivorous, the red fox eats small mammals, birds and bird eggs, frogs, insects, and berries. In hard times the fox will raid a hen house. Though the farmer may not agree, the fox's value in rodent control offsets the occasional loss of a chicken.

Red foxes are believed to mate for life. In late winter a pair digs a new den or readies an old one—or makes a home in a crevice or cave.

Early spring brings a litter of about five. Both parents help raise the young. In the words of zoologist Ernest Walker, "There is no more engaging sight than a litter of fox cubs tumbling and playing about the den, while being watched over by a parent." In early autumn, the family disperses, and each of the young claims its own territory.

OVERLEAF: *Teeth bared, red foxes square off over a meal. Blood is rarely shed in ritualized fights, in which dominance is the big issue.*

Kit Fox (*Vulpes macrotis*)

This perky little fox (here a pup nine weeks old) clearly deserves the name *macrotis,* long-eared. One of its nicknames, desert fox, describes its distribution. Sandy in color, its tail tipped with black, the kit fox dens in loose desert soil. It may either dig its own burrow or borrow one vacated by badger or prairie dog. For quick entry, its refuge may have as many as seven portals.

The kit fox is a small, slender animal. It rarely weighs more than 7 pounds (3.2 kg) or measures more than 32 inches (81 cm) long. The disproportionately large ears are acute sensors of chance or peril in the open desert country. Alerted, the kit fox readily responds. Over short distances it can usually outpace a predator, reaching speeds up to 25 miles (40 km) an hour.

V. macrotis hunts by night. Using a stalk-and-pounce strategy, it feeds on mice, rats, ground squirrels, birds, lizards, snakes, and insects. Eagles and coyotes prey on it. Ranchers pose an inadvertent threat. Poisoned baits laid out for coyotes are often taken by this canid. Once the kit fox was seriously threatened, but tighter controls on the use of poisons have helped bring it back.

Kit foxes mate in early winter and produce a litter of four to seven in early spring. While the vixen nurses the young, the male provides for her. In two or three months, both parents take the young out to hunt, and the family disperses in autumn.

Swift Fox (*Vulpes velox*)

The swift fox is well suited to its habitat, the high plains of North America. Here the tawny grasses cloak the buff-colored fox. Over flat, open terrain the swift fox can outrun all but the fastest predators. This aptly named fox is as fast as the kit fox, which it resembles in size and weight, as well as general body characteristics. Some mammalogists consider both foxes one species. But smaller ears, a broader skull, and its grassland distribution set the swift fox apart. In western Texas and eastern New Mexico where the ranges of the two species overlap, hybridization may occur.

These foxes breed from December to February. Gestation lasts 50 days and culminates with the birth of two to seven young. The pups emerge from their underground den at three weeks. Soon after, the parents begin showing them how to hunt. The family bond seems strong. If the female dies, the male will raise the pups alone. The young will stay with their parents until the fall.

Small mammals such as rabbits, rats, mice, and squirrels make up most of this night hunter's diet. It also eats birds, insects, lizards, and vegetation.

During the late 19th and early 20th centuries, this species declined drastically. It often took

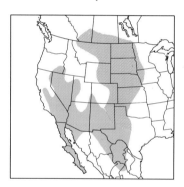

☐ Kit Fox
☐ Swift Fox

poisoned bait intended for other animals. The almost total conversion of the prairie to cropland destroyed much of the fox's habitat. The species is protected in some states and has been reintroduced in Canada, but its status remains precarious.

Gray Fox (*Urocyon cinereoargenteus*)

A gray fox speaks with a canid bark, hoarse and loud. It also has sharp, curved claws and catlike agility—it can scurry up a tree before most people can pronounce the string of Latin syllables that mean, loosely, "gray fox." Silvery gray from long snout to bushy tail topped and tipped with black, the tree-climbing fox measures 32 to 45 inches (81 to 114 cm).

Brushy country, forest, and rocky land suit the gray fox. Rather than dig its own den, this fox may occupy some natural crevice in rocks, a hollow log, or a den left vacant by a red fox or a groundhog. Pups average four to a litter, born in spring. Both parents care for them until they become able foragers.

Not finicky about food, grays take most anything rabbit-size or smaller: rats, birds, lizards, frogs, carrion, corn, nuts, berries, grasshoppers. Even centipedes and scorpions—stingers and all.

Spring flowers garland a gray fox nursery. The pups, about 7 weeks old, will help forage at 12 weeks.

Family Ursidae

The largest carnivores that walk the earth belong to the bear family. One massive male brown bear in coastal Alaska weighed 1,719 pounds (780 kg), a record in the wild. Polar bears rival brown bears in size. There may be a new heavyweight champion roaming a remote corner of the north. But weigh-ins are not part of a polar bear's routine. Polar bear cubs have a head start at birth—24 ounces (680 g) to the brown's 14 (397 g).

The ursid family numbers eight species, three North American: polar bear, black (opposite), and brown, or grizzly. At the subspecies level, and even among individuals of the same subspecies, bears vary so widely in skeletal characteristics that they have been hard to classify. Some bear specialists accept nine brown bear subspecies, up to 18 black bear subspecies, and one polar bear race for all of North America.

Bears are stocky, often slow-moving animals with low reproductive rates and overall low population densities. They are intelligent creatures with many learned abilities to utilize food and other resources in a wide variety of habitats. Because their food is only seasonally available, and because most bears den for long periods in winter, they have voracious appetites and must feed almost continuously when their food is abundant. Though diets vary greatly from place to place, vegetation is the mainstay for the black and the brown bears.

Polar bears depend primarily on the ringed seal for food but also eat plants and fish. Bears seek foods high in protein and sugars and are able to extract protein from plants almost as efficiently as herbivores. They can consume huge meals, up to 90 pounds (40 kg) a day for the brown and polar bears. And all three species can gain up to 7 pounds (3 kg) in a day.

The eyes and ears of bears, small in relation to their large heads, actually are far keener than most people suppose. Bears also have an acute sense of smell and, in their roaming, often pause to sniff the air—which perhaps accounts for the myth about weak eyes. Bears' lips are free from their gums, an asset in picking berries. Their teeth do not cut and tear as well as those of animals that eat meat exclusively.

Bears have five digits on each foot and have a plantigrade walk: They place the whole foot on the ground. They have strong claws suited to digging out food and dens.

All three species breed in spring, and all have delayed implantation of six to seven months. In late autumn, the partially developed egg implants and final growth starts, perhaps as the female enters her winter den. Cubs are tiny and helpless when born in midwinter and need much maternal care. Fiercely protective of their young, mothers also are strict disciplinarians, teaching their cubs feeding habits, orientation, and fear of other bears and humans.

Because of their dominance in the wildlife hierarchy, and their generally omnivorous food habits, bears competed with native North Americans and with European settlers for both space and food. And because bears are unpredictable, they posed threats—real and imagined. So people have eradicated them from vast areas of the continent.

Today, with growing numbers of people and shrinking habitats for bears, the competition is more intense than ever. And, though bears are not normally aggressive toward humans, they occasionally do hurt and kill people. Then more bears die.

Long-term studies of bear biology and behavior are vital to holding populations at healthy, productive, yet tolerable levels. Bears can adapt to people and to habitat changes people cause, but they may be near the limit of adaptability in many areas.

These truly wild animals need all the respect, attention—and living space—that sound management can provide.

Black Bear *(Ursus americanus)*

This is the "cute" bear, the teddy, the traffic jammer of national parks from the Great Smokies to the Sierra Nevada . . . the "car clouter" of Yosemite that bashes into cars to get food . . . the "mugger" of the Smokies backcountry that stalks hikers until their nerves snap and they drop their packs and run. This is the one that has dozens of researchers studying how to make the land safe for it—and for the humans whose habits have turned naturally shy bears into muggers, panhandlers, and dump addicts.

Despite human pressures, the black bear still has one of the most extensive ranges of any big animal on the continent. Scientists don't agree on how many subspecies have evolved, but color phases alone suggest numerous gene pools. Some eight to ten subspecies roam the Pacific edge, varying in color from totally black to the

pure white Kermodes bear of British Columbia.

In contrast to grizzlies, black bears climb trees easily, pushing off on hind legs, holding with forepaws. They come down in reverse, tail first. They may not look speedy, but they can sprint up to 35 miles (56 km) an hour.

When fresh scats vanish in bear country, it's a sign the winter sleep has begun. Winter denning varies with the length and severity of the season: six to seven months in Alaska, sometimes not at all in Mexico. Bears den under windfalls, in caves, or in tree holes—up to 60 feet (18 m) high in the Smokies. Often they use grass and moss to line their beds and to open and close the entrance for temperature

control. A denning bear is said to be dormant; metabolism rate and body temperature do not change sharply, as in true hibernation.

Half-pound (227-g) cubs— usually two, but as many as five— are born in the den. (The triplets above are two months old.) As adults they'll average 300 pounds (136 kg) for males, half that for females.

Black bears are not finicky eaters. Occasionally they rear up on hind legs to sift the breeze for food odors. Carrion? Delicious! Ants will do, or grasshoppers, grasses, acorns, fruits, fish, small mammals, even birds. As climbers, the bears do not have to wait for acorns to drop. In the Smokies they thrash through the tall oaks, leaving the woodland looking as if it had been battered by a storm.

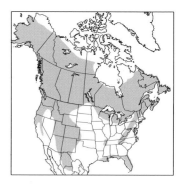

Grizzly Bear (*Ursus arctos*)

"Grizzly" means "grayish" and also "inspiring horror." Both meanings apply. The grizzly's thick, coarse fur varies in color—off-white, tan, yellow, brown, black. In the Rockies the typical hue is dark brown with a grizzly frosting on the back, source of the nickname "silvertip." It is also called the brown bear.

Naturalist George Ord put the second meaning of grizzly into a scientific name (*horribilis*) after reading of Lewis and Clark's adventures with this "tremendous looking animal." For years *Ursus horribilis* was classed as a North American species; now it is considered a race of the circumpolar brown bear, *U. arctos.*

Today authorities classify our grizzlies and mainland brown bears as one subspecies, *U. a. horribilis.* Another race, *U. a. middendorffi,* called the Kodiak bear, inhabits Kodiak and nearby islands in the Gulf of Alaska.

Grizzlies average about twice the weight of the black bear, weighing 600 to 800 pounds (272 to 363 kg) as adult males. But size may not offer a good clue to the identity of a lone bear spotted on a distant trail. Where does the bear loom tallest? At the shoulders? The hump of muscle there identifies the grizzly. Farther back, toward the rump? Then it's a black bear.

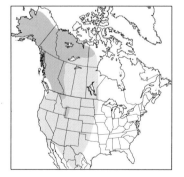

When is a grizzly not a grizzly? When it's coastal—in Alaska. Common usage there applies "grizzly" to inland bears. Those that roam the coasts and islands are called Alaskan brown bears. Both display the humped shoulder, dished face, and long claws of their species. But at 1,600 pounds (725 kg) the beachcomber above may weigh a third more than the grizzly.

Present Range
Former Range

Grizzlies mate in late spring. Cubs, usually two weighing about 14 ounces (397 g), are born in the winter den; they stay with mother some 18 months. She becomes sexually active as contact within the family group declines and she leaves the cubs. Or she—or her mate—may even run them off.

Roots, leaves, and berries form the bulk of the diet, but grizzlies also relish meat: squirrel, elk, moose, deer—freshly killed or carrion. They feed in garbage dumps and pay the price for eating humans' sugary food: tooth decay. At times they prey on cattle. They avoid humans—but not always, and with tragic results for both.

The grizzly has been eliminated from parts of Canada, Mexico, and the United States. It bestrides the flag of California but is gone from there. South of Canada it has some protection as a threatened species. Even so, it is often shot as a threat to people and livestock. With habitat loss and the growing

human presence in the northern Rockies, grizzly survival even in national parks depends upon research and wise management.

Yet the grizzly remains a force, a symbol of untrammeled nature: "His is a dignity and power," wrote outdoorsman Andy Russell, "matched by no other in the . . . wilderness."

A paw flicks out, pins a fish to the riverbed; jaws snap, securing the catch. Here at McNeil Falls at the base of the Alaska Peninsula, salmon swim upriver to spawn in July and August. They mass at the rapids, as do brown bears of all ages and fishing skills—up to 60 or 80 of them along a 100-yard (91-m) stretch.

Brown bears are the most social of North American ursids. They tend to congregate at food sources and often form family foraging groups with more than one age class of young. Social hierarchies develop as the bears arrive. Biggest males get the best fishing sites. Next come she-bears with cubs, then females without young, sibling groups that stick together, and, last, the small loners.

To make the most of limited fishing space and time at McNeil Falls, the brown bears evolved a stable society that avoids useless combat. A massive, battle-scarred male need not prove himself every time out; the sight of him spooks younger, smaller bears—and he can concentrate on the fish. Dominance bouts often end with open-mouth threats, but not always. One huge male was seen to bite into a 750-pound (340-km) rival and lift him off the ground. The bloodied loser returned the very next day.

Young bears, such as the ones sparring on these pages, seem to spend their prime time in social play, though roughhousing may lead to a serious scrap.

Protective females start the most fights. For all their closeness, cubs and mothers often get confused about which belongs to which. Several litters may follow a single female; she'll nurse the lot—then let the adoptees choose whether they'll return to their real mother.

With late summer, berry crops lure the brown bears from the river. The social order dissolves, and it's back to the rover's life.

Polar Bear (*Ursus maritimus*)

Creature of snow and ice and frigid seas, the polar bear spends much of its time adrift, stalking seals on pack ice and floes. Its white coat— and a body adapted to water— make it seem unique. Actually it is closely related to the brown bear. Mating between the species in zoos has produced fertile offspring, confirming the kinship.

Sizes and weights of the largest polar bears approximate those of the largest brown bears. Adult males may weigh more than 1,200 pounds (544 kg). But they usually are 8 to 11 feet long (2.4 to 3.4 m) and weigh 900 to 1,100 pounds (400 to 500 kg). Females weigh less than half that, and both males and

females are noticeably smaller in the high Arctic than at the southern limits of their range. The polar bear does not have the brown bear's scooped face.

The polar bear's coat, excellent as insulation against the cold, camouflages its hunting activities— and may channel ultraviolet light to its skin. Hairs examined under

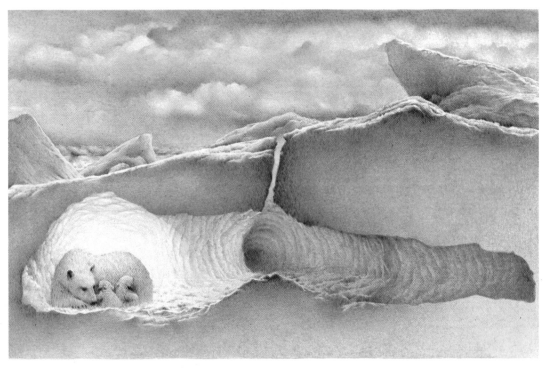

A three-month-old cub sticks to its mother in their snowy realm. Polar bears use traditional maternity denning areas where as many as 200 females may gather.

Typically, each digs her den in a snowbank, entering it in October and adding a chamber as winter drifts deepen. A vent to the surface admits fresh air.

Lethargic during her lying-in, the mother responds with cuddling to the cries of her cubs—usually twins, born November to the end of January. Through the lair's thin dome, increasing light cues the approach of spring. The mother breaks through the ceiling and emerges, giving her progeny their first peek at the outside world.

electron microscopes show up as transparent tubes with inner surfaces that reflect light. Thus the white fur may enhance heat absorption. The bear's eye has a well-developed nictitating membrane that acts as a shield to protect the eye from stinging snow and the sun's blinding glare. Polar bears roam far in search of food—

but they are not aimless nomads. Their seasonal movements follow definite patterns. Deported garbage-dump raiders unerringly made their way hundreds of miles back to Cape Churchill. This most carnivorous of the bears subsists in summer on plant foods, birds, and other flesh—including (above at left) a stranded whale.

In times past, the deep-freeze domain of the ice bear seemed safe from mechanized intrusion. There was a saying: "A polar bear sees a man once in its life—when it is shot." It no longer applies. The race to extract oil and gas from the polar seas encroaches on its range, and concern for this species grows.

OVERLEAF: *Monarchs of the Arctic wilds, a female and near-grown cubs clamber onto an ice raft. In their second winter cubs hunt seals with their mothers on the pack ice, denning only during severe storms.*

Family Procyonidae

North America's procyonids—the raccoon, ringtail, and coati—have ringed tails like the one on Davy Crockett's coonskin cap. These species range from the size of a house cat to that of a medium-size dog.

They have extremely well-developed and useful front feet. Each foot has five digits. The animals walk on their toes and at least part of their soles.

Raccoons, ringtails, and coatis are chiefly nocturnal and are excellent climbers. They often rest in trees during the day. Offspring usually are born in tree-trunk cavities. All members of the family are believed to be promiscuous breeders, with males playing little or no role in providing for the young.

Procyonids will eat almost anything: eggs, mice, berries, crayfish, fish, corn, insects, spiders, young rabbits, nuts, fruits. On their nightly forays, they explore just about every nook within their reach. Their dexterity and natural inquisitiveness often lead them to a tasty snack.

Ringtail (*Bassariscus astutus*)

Late at night, after raccoons and striped skunks have eaten their fill and moved on, the ringtail creeps out of its rock crevice to forage. The ringtail has much to choose from in its habitat of desert canyon ledges or brushy slopes. In winter it eats ground squirrels and smaller rodents; in summer, insects, spiders, and desert plants. For this expert climber even the contents of a bird's nest are within easy reach. With a hind foot that swivels 180°, the ringtail ranks among the most dexterous of mammals. In recent years it has extended its range northeastward into Kansas, Arkansas, and Louisiana.

The ringtail resembles many animals, but none closely. Its scientific name means "clever little fox"—its face is foxlike. Its slender body, small feet, and a genius for catching mice gave it many common names ending in "cat." One, miner's cat, dates from gold-rush days when prospectors tamed ringtails for company—and to rid their camps of mice.

The big-eyed, buff-colored ringtail resembles the raccoon most in its bushy, black-banded tail. An organ of balance, it measures up to 17 inches (43 cm), half the ringtail's overall length. And, arched over the animal's back, it makes the little "cat" appear formidable to possible predators.

Strictly nocturnal and quite shy, the ringtail is rarely seen, though it is not uncommon. Litters of three or four young are born in spring.

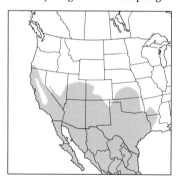

Coati *(Nasua nasua)*

The coati's tough, sensitive, and flexible snout has given it at least one nickname: hog-nosed coon. Unlike other procyonids, this one is most active during the day. It is omnivorous—feeding on anything from berries and mice to lizards and insects. (The mother and daughter above examine the prospects for an earthworm dig.)

A long, upright tail measuring up to 27 inches (69 cm) equals the coati's body length and often leads people to mistake a coati for a monkey. But this tree climber's tail, though helpful as a balance, cannot be used to grasp.

The rust-colored coati has a dusky face mask and small, soft brown eyes that give it a wistful look. The most gregarious of the procyonids, coatis form casual bands of 4 to 30 or more females and young that live and forage together. An adult male from the group's home range is welcome to join only during the April mating season. Litters of two or more offspring are born in early summer.

The coatis' range reaches from woodlands in Central America and Mexico into the southwestern United States. They crossed the border into Texas about 1900, perhaps as their predators were exterminated.

Raccoon *(Procyon lotor)*

The raccoon, equally at home in city, suburb, or trackless forest, is among the few mammals that thrive in the face of encroaching civilization. It is probably more numerous now in the United States than it was when Captain John Smith explored the New World. It is Canada's only procyonid.

Its diminutive tracks, resembling a human hand in shape, may lead from a hollow tree into mischief. Raccoons are notorious for their nighttime raids on garbage cans. Prying off the lids is no challenge for an animal with thin, mobile

fingers and a fine sense of touch. Raccoons kept as pets have learned to turn on faucets, open latches, turn knobs, and manipulate other devices that get them into trouble.

Captive raccoons often douse and "wash" their food in water—a habit that has earned them their Latin name, *lotor,* washer. No one knows why they do this, but cleanliness is not the reason. They will also wash food—or just their paws—when away from water.

Raccoons measure about 32 inches (81 cm), including tails, and usually weigh about 20 pounds (9 kg). Their litters of 2 to 7 young,

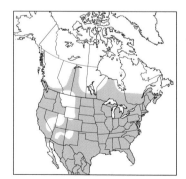

born in the spring, are reared by the female. A choice den site is a hollow tree 20 to 40 feet (6 to 12 m) up, but any shelter will do.

In winter, northern raccoons become dormant but do not hibernate. A Minnesota trapper entered an old cabin one winter, shined his light about, and beheld a startling sight: 23 pairs of beady eyes steadily staring at him.

Four young raccoons find a hollow tree ideal for playing follow-the-leader. Sprightly and mischievous, they rank among the most playful of mammals.

Young raccoons born in a tree have to climb to the ground before they begin to walk around. How do they do it? Any way they want to—head first or tail first or, as at left, using a two-trunk system. It helps to have hind feet that, if need be, can rotate a full 45°.

Raccoons get plenty of climbing practice. High branches are fine places to sunbathe and to sleep when the raccoon isn't hungry. And when it is, those remarkable paws are good for more than just climbing. They

gave the raccoon its common name. "There is a beast they call Aroughcun," wrote Captain John Smith in 1612 of this New World creature, prized by the Indians for its flesh and its fur. Aroughcun, or raccoon, means "he scratches with his hands." Seeming never to pause, those sensitive hands scratch, explore, feel, and poke around for food. Rotting tree stumps are likely places to prospect for termites and ant larvae. No stone, crevice, or even hornet's nest escapes scrutiny. Thick fur protects the coon from stings. Omnivorous, it gains weight rapidly when food is plentiful, storing up fat for the winter.

Swimmers as well as climbers, raccoons can get to delicacies—such as fish, frogs, clams, and crayfish— not accessible to more firmly grounded creatures.

Family Mustelidae

The weasel family is made up of the most diverse species of all the carnivores. It includes badgers and otters, minks and skunks, weasels and fishers. With their differing lifestyles, one or more of the mustelids has adapted to virtually every type of terrestrial habitat—undergound to treetop—and to aquatic environments, both fresh and salt water.

All have evolved the strong, sharp teeth typical of carnivores. Dependent upon variable and uncertain food resources, they pursue what they need with intelligence and in constantly changing ways. This family, along with canids, felids, and others arose from the miacids, primitive carnivores of 55 million years ago.

Most mustelids are small, with long, slender bodies and dished-in faces. They vary from the cigar-size least weasel to the sea otter, which may weigh up to 100 pounds (45 kg). Their motions are quick if not always graceful, and they possess prodigious strength and endurance. Their legs are short and powerful, and their five-toed feet are adapted for running, digging, climbing, or swimming. Some mustelids walk on their toes, others on their soles as well. Claws are not retractile.

Most mustelids possess anal scent glands, which are used for defense and for marking territory. These glands are best developed in skunks, which rely on scent to ward off enemies. The mechanism is so potent and memorable that skunks often make no attempt to escape enemies. For most carnivores, it is advantageous to blend with the surroundings. But skunks advertise. Their coloration may remind foes that they will be sprayed—and forestalls attack. Peaceable, usually in harmony with human aims, the skunk plods about unmolested.

But the wolverine is a mustelid whose behavior and temperament stir enmity whenever it crosses paths with people.

Raiding traplines, wreaking havoc, the wily wolverine has been a competitor to be warred upon and destroyed. Extraordinarily strong, it is perhaps the most tenacious member of a tenacious family. But attitudes toward this wild spirit of the north have changed in parts of its range, and it has been protected in some areas.

Conversion of vast areas of the continent to farms, highway complexes, reservoirs, cities, and wasteland has harmed most mustelid species, but not all. Skunks profit, for instance, where the pine marten and fisher—forest-loving animals—lose. The badger gains ground where farmland is marginal or reverts to a semiwild state. The black-footed ferret, a heavy loser, has been all but eliminated.

Many mustelids have the misfortune of possessing very desirable coats. Exploitation of the furbearers varies in intensity as market values of furs rise and fall with the vagaries of fashion, and mustelid population levels are affected accordingly. Trapping has wiped out some species in parts of their range and jeopardized others.

The reproductive potential of mustelids is unusual and widely varied. In most, implantation of the fertilized egg, or blastocyst, is delayed, extending pregnancy to a time of year favorable for birth and survival of young. In the fisher, the result is almost a year of pregnancy. But in the mink, birth takes place about two months after mating. In both species, growth of the young in the womb is of short duration, and they are born poorly developed. Litter sizes in mustelids cover a wide range, from as many as ten for the ermine to only one for the sea otter of the Pacific coast.

That bewhiskered mustelid (placidly dining on sea urchins opposite) was hunted to the brink of extinction. Given protection, it has made a slow, solid recovery in one of the great conservation success stories.

Marten *(Martes americana)*

Neither deep snow nor the numbing cold of northern winter keeps the marten from its rounds. Frisking over the drifts on furred feet that keep it from sinking in, tunneling beneath when the signs are right, it pokes into every deadfall or sheltered cranny that promises a meal of vole, chipmunk, mouse, hare, or shrew. If action stirs aloft, up it scampers and a branch-to-branch chase ensues. In a marten-versus-squirrel contest, the carnivore is the odds-on favorite.

Among the mustelids, martens are much more carnivorous than skunks and less so than weasels. Birds' eggs and insects vary the

marten's summer diet. When blueberries are ripe, the marten gets blue lips.

Mature evergreen forests provide habitats suited to this solitary species, also called pine marten or American sable. Each adult animal, the size of a house cat, may need a home range of 15 square miles (38 sq km) in times of food scarcity; in times of plenty, only a fraction of that. In addition to the anal scent glands of most mustelids, martens have on their bellies scent glands that are rubbed against the ground to mark territory or to signal the onset of mating season. Females in estrus also use a vocal signal: They cluck. After potential mates meet, a reluctant female may be dragged by the scruff of the neck until she

cooperates. Her young number two to four and arrive in early spring.

Loss of habitat from logging, fire, and spreading humanity has robbed the wilderness-loving marten of many former homes.

Its intense curiosity makes this beautifully furred animal an easy mark for trappers. Almost any bait, edible or not, will work. And some martens never do learn. Biologists studying the species in Glacier National Park live-trapped one hyperinquisitive male 77 times.

The snap of a trap is usually the sad end of a tale—but not always. Some martens taken in traps of a nonlethal sort are being transplanted to logged-over forests that have regrown—in the hope new populations will flourish.

Fisher *(Martes pennanti)*

The fisher does not fish for a living. Instead, this close relative of the marten is known for its speed through the treetops. Wrote naturalist Ernest Thompson Seton, "It is probably our most active arboreal animal. The squirrel is considered a marvel of nimbleness, but the marten can catch the squirrel, and the fisher can catch the marten."

A fisher sometimes does catch and eat a marten, but more often it feeds on hares, small rodents, birds, carrion, and fruit. Where its range overlaps the marten's and they compete for food, the two may specialize by size, the larger fisher taking larger prey.

With a unique degree of success the fisher makes a food resource of the well-armed and populous porcupine. For a long time it was thought the fisher's technique involved upending the lumbering rodent and attacking the flip side, which lacks quills. Not so, studies show. The porcupine's defense works best against attack from above and behind. The fisher's low profile puts it at the right level for a frontal attack, and it uses teeth and claws to inflict wounds on the porcupine's face and neck before tearing into its belly.

Because too many porcupines can debark and kill too many trees, foresters and fishers have become allies in parts of former fisher habitat—in Montana, Idaho, Oregon, Vermont, Michigan, Wisconsin, and West Virginia. Finding a few hungry fishers more desirable than buckets of poison,

wildlife managers have restocked the deep-woods predator to kill off excess porcupines and save trees.

At home in the wilderness, the fisher finds a hollow tree or log a fit den for its slim body—up to 25 inches (64 cm) long, excluding the tail. The one to five young are born in March or April and become independent by fall.

A female fisher mates a week or so after bearing young, then undergoes a gestation period of up to 358 days. Of this, ten months pass before the embryo becomes implanted; its actual development takes about two months. Thus, in a sense, the female spends nearly all her adult life in a state of pregnancy.

Ermine *(Mustela erminea)*

Long, slender skulls and sinuous bodies fit the weasels for their niche in life: They are a scourge of small burrowers, such as mice, voles, moles, and chipmunks. Weasels can enter any burrow or hole they can get their heads into.

In the open, a weasel stalks chiefly by scent. In lightninglike moves it pounces on its prey with clawed forelegs and kills it with bites to the back of the neck. Weasels can kill rats, rabbits, and squirrels larger than themselves.

Middle-size among the three weasel species in North America, the ermine weighs 1.6 to 3.7 ounces (45 to 105 g) and is 7.5 to 13 inches (19 to 34 cm) long. About a third is tail, giving the ermine another common name, short-tailed weasel (though the least weasel's tail is even shorter). The ermine in North America

ranges from northern states to above the Arctic Circle. Weasels are bolder than their small size warrants. On occasion they have attacked humans who stood between them and their food.

Captive weasels eat a third or more of their weight each day. Seldom does nature provide a steady source of food. So the weasel's mode of survival is to kill whatever it can whenever it can and store the surplus. Its den often has a side tunnel used as a storeroom for slain mice. Confronted with an unnatural surfeit of food, as in a farmer's hen house, the voracious ermine follows the only pattern it knows— and overkills.

The ermine breeds in early summer. After delayed implantation, four to ten young are born in the spring.

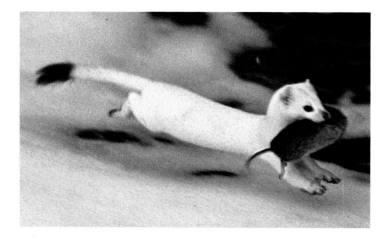

Ermine in winter pelage spirits away a vole. Brown in summer, the weasel wears white in winter. Blending with the season, it benefits as hunter and hunted. The tip of its tail remains dark.

A predator, seeing only the spot of color moving against the snow, may strike behind it.

Long-tailed Weasel (*Mustela frenata*)

This largest of North American weasels weighs 2.5 to 9.4 ounces (72 to 267 g) and is 12 to 22 inches (30 to 55 cm) long, including a 4-to 6-inch (10- to 15-cm) tail. Found from Peru to Canada, it has a range that overlaps those of the ermine and the least weasel.

The presence of three similar carnivores in an area, feeding on essentially the same prey, is unusual. But this weasel's size enables it to kill larger prey, thus helping assure all three species adequate food. Coyotes, foxes, hawks, and owls that prey on all three weasels reduce the populations of each and ease competition for the same food.

In the northern part of its range the longtail, like the other two weasels, changes color from dark brown to white in autumn. (Pelage of the ermine—opposite, top—is in the process of change.) As the days grow shorter, less light enters the weasel's body through its eyes, stimulating molt by means of the pituitary. This gland also inhibits release of pigment to the cells in the hair follicles, resulting in a color change as the pelage regrows. A second molt as the days grow longer reverses the color scheme.

As persistent and fearless a hunter as its fellow weasels, the longtail is also equally adept at climbing and swimming. It seems to prefer more open habitat.

Breeding in midsummer, the longtail bears three to nine young in the spring, implantation having been delayed about eight months. The new generation reaches maturity within six months, and females are able to mate in their first year of life.

Least Weasel (*Mustela nivalis*)

The world's smallest carnivore, the least weasel may weigh less than an ounce and hardly tops 2 ounces (28 to 57 g). It is at most 9.8 inches (25 cm) long, about a quarter of that tail (not black-tipped, as on other weasels).

A high metabolic rate drives the least weasel constantly in search of mice, its principal quarry. A rise or fall in mouse population may have a similar effect on weasel numbers. Like other weasels, it often usurps a victim's burrow and uses it as a base for more forays. To prepare the den as a nursery, the female may line it with the victim's fur.

Least weasels do not experience delayed implantation. Thus, they can produce two litters a year. Litter sizes vary from three to ten.

Speed, ferocity, and the ability to crawl into tight spaces help it cope with an array of predators. As a last resort, the least weasel can emit an odor said to be as pungent as that of the striped skunk, though it cannot spray the musk as skunks do.

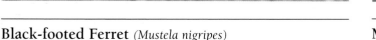

Black-footed Ferret (Mustela nigripes)

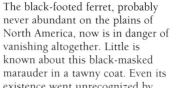

The black-footed ferret, probably never abundant on the plains of North America, now is in danger of vanishing altogether. Little is known about this black-masked marauder in a tawny coat. Even its existence went unrecognized by scientists until 1851, when John James Audubon and John Bachman described it from a skin provided by a trapper. Today it ranks among the continent's rarest mammals.

Active mainly at night and at dawn, the ferret depends heavily on the prairie dog for food and shelter. Originally, the ranges of the two species coincided: from Texas north across the grasslands into Alberta and Saskatchewan. But cattle ranchers, seeing the prairie dog as a competing grazer, launched massive campaigns to wipe it out. And in this war the ferret's habitat was disrupted.

Ferrets may also prey on mice, ground squirrels, and gophers. But prairie dogs are the ferret's chief food. The ferret hunts mostly underground, within the prairie dog's labyrinthine passageways. The ferret seizes a victim by the throat, kills it, and drags it out to the ferret's burrow. Prairie dogs show little or no fear of the ferret above ground. Below, they may try to keep a ferret in or out of a tunnel by plugging it with dirt—a strategy that apparently seldom succeeds.

To feed herself and an average litter of three, a female ferret would need, by one estimate, a larder consisting of a prairie-dog town covering at least 140 acres (56 ha).

Ferret young are born in the spring. In summer, the mother takes them above ground, usually at night. In the fall, the young disperse to lead solitary lives. At full growth, they will measure about 22 inches (56 cm) nose to tail, and weigh 2 or 3 pounds (0.9 or 1.4 kg).

Captive-bred animals are being used in reintroduction efforts.

Mink (Mustela vison)

Elegant in its lustrous brown coat, a mink slinks from its riverbank den. It was a muskrat's den, until the mink killed and ate the owner. Hungry again, always hungry, the mink embarks on another foray. All night and at times by day, in the water and on land, this close kin of the aggressive weasels prowls— alone. Except briefly at breeding time, a mink will not live in a place crowded with others of its kind, even though food is plentiful.

The mink climbs well and, having semiwebbed feet, swims well, the rich fur protecting its streamlined body from icy water. An amphibious mode of living puts great variety into the list of possible prey, including fish and crayfish. One food study in a North Dakota prairie marsh tallied 32 species of prey identified from mink scats and uneaten remains in and around their dens. Dabbling and diving ducks, blackbirds, frogs, and garter snakes were among the kills, as well as the voles, mice, and shrews that are weasel staples. The versatile mink may be found from Florida into the Arctic, in all but very dry areas.

The litter of two to ten kits, born in spring, receives tender care. The family may change dens several times before the kits mature and disperse. Minks measure up to 28 inches (71 cm), including tail, and weigh up to 3.5 pounds (1.6 kg).

Large owls, bobcats, coyotes, and human trappers prey on minks. Commercial ranches supply most of the demand for pelts. As many as 100 are needed to make one full-length coat.

Wolverine *(Gulo gulo)*

Legendary "devil beast" of tundra and boreal forest, the wolverine is probably unexcelled in strength by any mammal its size. This giant weasel, bearlike in shape, measures 36 to 44 inches (91 to 112 cm) from nose to tail tip. It weighs at most 70 pounds (32 kg). Yet it has chewed through and pulled down logs a foot (30 cm) thick to rob a trapper's food cache and has killed moose 20 times its size. Usually when an outsize animal becomes wolverine fare, it is old, weak, hampered by deep snow, or already dead. *G. gulo* will eat as much as it can hold of anything it can get.

Winter is a time of hunger for many arctic animals, but the wolverine does not migrate or hibernate. Mastery of snow travel aids in making a living. Broad, furred feet spread the weight of this formidable snow mammal, letting it lope over crusts. It may cover 30 miles (48 km) a day to find prey or scavenge kills of other predators, including humans. Wolverines are notoriously adept at working traplines. They not only eat the meat but also may carry away the traps. Occasionally, they scent-

Present Range
Former Range

brand food they have no immediate use for. The elusive wolverine is better known by its works and its odor than by sight.

In a den under the snow or other sheltered place, two to five young are born from January to April. They grow rapidly and are on their own by the following winter.

Traps and poisons, along with loss of habitat, almost wiped out the wolverine in the United States. Remnant populations persist in the Rockies of Montana and the Sierra Nevada of California.

Badger (*Taxidea taxus*)

The scene of a badger's life story keeps shifting from one hole in the ground to another. Born and bred in a burrow, it spends a good part of its time digging in and crawling out. In summer it may go through a den a day, not sleeping in the same place two days in succession. In cooler weather a longer tenancy is usual. If it digs up a big meal, such as a rabbit, the badger may drag its kill into a tunnel and hole up for several days. The usual fare is smaller—ground squirrels, mice, snakes, bees—and every day is the same old story: dig, dig, dig.

This resident of arid grasslands and sagebrush country digs not only to eat, rest, and nest, but also to bury its feces and to take refuge from belligerent farm dogs.

Badgers have a body shaped for life underground—short and flat. The animal can flatten its body to wriggle and crawl through a network of burrows. Its legs are short. Its long, strong claws are designed for digging. These, combined with the badger's ability to shove loose earth with its body, make it the most prodigious burrower among carnivores. It measures up to 34 inches (86 cm), including tail, and weighs up to 22 pounds (10 kg).

It may live close to humans but shuns contact with them. Its burrows can be a problem on farms, and cowboys have always regarded badger holes as hazards for horses.

When cornered, badgers fight fiercely, and men once trapped and baited them for sport. They would put one in an open barrel or similar place and force it to fight against dogs. The badger would back into its corner and, growling and snarling, face its tormentors. From the cruel sport of baiting badgers came our usage of "badger"—to harass and worry.

An adult badger normally wanders alone except at mating time. When a female has her one to five cubs, she settles in one burrow for about a month, until the young can venture out with her and seek another home on the range—and another and another.

Striped Skunk (*Mephitis mephitis*)

Turning its back to the enemy, arching its spine, and lifting its tail, the striped skunk sprays an oily, foul-smelling fluid up to 15 feet (4.5 m) away. The odor can travel half a mile (0.8 km) on the wind.

Most mustelid species have well-developed scent glands. But in *M. mephitis* and other skunks, this trait has evolved into an important means of survival. Two glands located at the base of the tail contain about three teaspoonsful of musk, enough for five or six discharges. The skunk can replace

this fluid at a rate of about two teaspoonsful a week. Spraying is the ultimate response to danger, and warning signals usually precede its use: hisses, growls, foot stamping, tail waving. Though rarely used, the chemical defense is so effective that the skunk's name and reputation derive from it. *Mephitis* means "bad odor."

This creature is peaceable, plodding, and all but harmless if left alone. Black, with a distinctive forked stripe along its back, it is the most common skunk in North America. It occurs from Canada's Northwest Territories to central

Mexico. Preferring fields over forests, it has adapted well to human alterations of the land. Although it generally inhabits natural crevices or dens abandoned by other animals, it also lives under buildings, in woodpiles, or around garbage dumps. An omnivore, it eats vegetable foods, grubs, insects, mice, shrews, eggs, and carrion. In total length, males measure 23 to 31.5 inches (58 to 80 cm). Females are somewhat smaller.

A pleasant sight of late springtime is a skunk parade along the roadside, a mother with her four to seven kits following single file, out for an evening stroll and a meal. Used to getting the right-of-way, skunks have poor road sense, and traffic takes an odoriferous toll.

Striped skunks have few predators. Humans shoot or trap them. Bobcats attack on occasion. So does the great horned owl, which has better eyesight than sense of smell.

Hooded Skunk (*Mephitis macroura*)

Markedly hairy at each end, this skunk is well named—in two languages. Longer hairs around the neck resemble a hood; *macroura* means "long-tailed." That bushy appendage accounts for half the skunk's total length, 26 inches (66 cm). Some hooded skunks are almost totally black; others are well streaked with white.

This species lives in deserts of Mexico, southern Arizona and New Mexico, and western Texas. In streamside brushlands and rocky canyons it forages for insects—the mainstay of its diet—and small rodents, bird eggs, and plant foods. It serves as an important check on the populations of large insects and small rodents that thrive in the year-round mild climate of its homeland. Little is known about the hooded skunk; its manner of living is probably similar to that of the striped skunk. Its dens most often are found in rocky crevices or dug into the soil. Unlike the striped skunk, this one seldom lives in abandoned cabins or under sheds.

One litter is born each year, probably with three to five young. The species is nocturnal and, except for an occasional coyote or bobcat, is seldom preyed upon.

Hog-nosed Skunk (*Conepatus mesoleucus*)

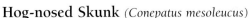

With its bare, flexible snout the hog-nosed skunk roots for insects, grubs, and worms in desert valleys and brushy canyons of Mexico and the southwestern United States. Sturdy claws on its forelegs aid the quest. Its diggings look much like the work of hogs. The hog-nosed skunk also eats small rodents, reptiles, and vegetation, including prickly-pear cactus fruit.

Little is known of this species' reproductive behavior. A litter of two to four young is born in spring after a gestation period of about two months. The adult hognose weighs between 3.4 and 10 pounds (1.5 to 4.5 kg) and measures 23 to 27 inches (58 to 69 cm).

Nocturnal and solitary, it is seldom seen foraging. When met, it pays an onlooker little heed. Head down, intent on rummaging through the soil for food, it seems to fear no predator. Trappers do not value its coarse, brown-tinged fur. Its musk can send even rattlesnakes wriggling away.

Conepatus, little fox, is South America's only skunk genus, and only *C. mesoleucus* ranges into the United States. It reaches as far north as Colorado, where it is never abundant. The hognose is threatened by civilization's advance into the desert, especially into less arid canyon bottoms, where loamy soil supports enough insects to sustain the skunk.

Spotted Skunk (*Spilogale putorius*)

The smallest North American skunk is distinguished by its coat, a silky collage of spots and broken white stripes. Adults are about 20 inches long (51 cm) and weigh about 2 pounds (0.9 kg). Spotted skunks are adept at clambering up trees—and occasionally live in them. More often, they den underground. Small size and the ability to climb enable this species to steal into hen houses more easily than other skunks. Though fond of eggs, they usually eat rodents, insects, snakes, and fruit. Owls, foxes, and bobcats sometimes prey on them.

Spilogale putorius—roughly translated, "spotted stinker"— ranges from British Columbia and the Appalachian Mountains to Central America. The female bears a litter of two to six in the spring.

"Fair warning" says the artful handstand and waving tail of a spotted skunk. If the enemy comes closer, it gets sprayed.

River Otter (*Lutra canadensis*)

The engaging river otter once cavorted in and out of water across most of the United States and Canada, appearing to enjoy life thoroughly. It can live near people and seems to like showing off. But it has been trapped, shot, poisoned, and polluted to death in most of the central and southwestern U. S. The otter's lithe, streamlined body, with short legs and webbed feet, enables the animal to swim at speeds reaching 7 miles (11 km) an hour. Adults are as long as 51 inches (130 cm), including the fleshy, tapered tail that serves as a prop on land, a rudder or oar in water.

Otters mate in the water, usually in winter or early spring. Male and female then go their separate ways. One to five pups are born nearly a year later in a riverbank den the female prepares. At about 12 weeks the young venture out. Soon they are hunting their favorite foods—fish, crayfish, frogs, insects, and small mammals.

Rhythmic as a ripple, the otter dives and surfaces, undulating through the water by flexing its body up and down. Watch a "train" of otters snake down the middle of a river and you may think you're seeing a sea serpent of monstrous size. (You won't be the first.) This adept aquanaut can easily dive to 35 feet (10 m). Flaps of skin close its nose and ears and its pulse rate slows, allowing the animal two minutes underwater before it must pop up for air.

Agility and speed in the water come in handy not only for catching fish but also for catching pebbles tossed in a one-otter game of ball— or for strategic withdrawal after a teasing tug on a beaver's tail.

Body-sledding down a snowbank is another favorite activity.
Otters go as a family group. Adults take turns with the young at tobogganing on their bellies.
When there is no snow, a slippery mudbank will do, preferably angled to the river so the slide can end in a resounding belly flop.

Sea Otter (*Enhydra lutris*)

Lolling on a kelp bed along the Pacific coast, shielded from frigid water by luxurious fur, the sea otter seems to lead an easy life.

Its ancestors once lived on land. After taking to the sea eons ago, they did not develop a blubbery layer beneath the skin, as whales did. The otter depends for protection from the cold on the blanket of air trapped in its densely packed fur, a fur so fine it almost doomed the species.

Said Captain James Cook after acquiring some pelts from Nootka Indians in 1778: "The fur of these animals . . . is certainly softer and finer than that of any others we know of." Sea otters were already being killed for their pelts by Europeans, Asians, and North Americans. Cook, China-bound, took furs with him. The demand and the slaughter grew. The fur trade nearly wiped out the species.

In 1911 the United States, Great Britain, Russia, and Japan agreed to stop the killing. This near-shore animal has made a substantial comeback in the Aleutian Islands and off the California coast south of Monterey. It shows promise where it has been transplanted.

The sea otter, 4 to 6 feet long (122 to 183 cm), usually weighs 33 to 66 pounds (15 to 30 kg), but it may reach 100 pounds (45 kg). The male is the largest North American mustelid. Females are about 20 percent smaller.

Adults first breed at about four years, courting and mating in the water. A single pup—rarely two—is born six to eight months later. With no margin for error in a litter of one, the newborn is better developed than most mustelid pups, arriving eyes-open with a mouthful of milk teeth. For a year the pup will nurse, nap, and be groomed. Its mother will carry it on her chest while she floats or swims on her back. Males usually live apart.

Grooming is not a mere nicety. If the otter's coat—containing some 800 million fibers—gets soiled or matted, the trapped air is lost and with it buoyancy and insulation. Oil spills and other pollution—and competition with commercial fishermen for some of its favorite foods—are among the problems that still menace the otter.

A dive to the bottom puts a large abalone within this successful hunter's grasp (left). Sea otters bring their food to the surface to eat, hauling it up in pouchlike chest folds or clasping it with their paws. They may dive as deep as 180 feet (55 m) to find edibles—chiefly mollusks, crabs, and sea urchins—and may eat a fifth of their body weight in food each day. The otter above uses a rock balanced on its chest as an anvil to batter open a clam.

Family Felidae

The cats are the most lithe and graceful of all the mammals. And, of all the carnivores, cats are perhaps the most proficient killers. Members of this family possess keen senses and lean, muscular bodies ideally adapted for a predatory existence. Forelimbs are armed with sharp claws. Longer hind limbs provide for amazing leaps and bursts of speed. Jaws

Cats bare five long, curved claws to grasp prey or rip at an enemy. When not in use, the talons retract into sheaths to protect the sharp points. Furred feet of the felids permit a soft and noiseless tread.

and teeth are designed to seize, kill, and devour other animals. Some cat species are capable of killing prey far larger than themselves. The felids are among nature's finest physical machines.

Taxonomically, the cats of the world are divided into *Panthera,* the large roaring cats, and *Felis,* the smaller purring cats. Because of the different way the voice box is attached, the members of *Panthera* rasp or roar in deep, gruff tones; members of *Felis* purr continuously or make shrill, higher pitched sounds. Of the seven cat species found in North America, only the jaguar belongs to the genus *Panthera.* The other six—mountain lion (or cougar), lynx, bobcat, margay, ocelot, and jaguarundi—are purring cats and are

members of *Felis,* although many modern authorities assign the bobcat and lynx to a third genus, *Lynx.*

In both form and function cats reach the peak of predatory evolution. The teeth, fewest in number among carnivore families, are also the most specialized. The canines are better developed for seizing and perforating prey than those of any other flesh eater. Even the tongue is specialized. It is covered with sharp, horny protuberances that rasp meat from bones. The tongue also is effective in grooming the animal's coat, which is one reason cats look so sleek.

Outwardly, all cats appear very "catlike." Their coat colors differ. The jaguarundi and cougar (opposite) have plain coats; others are spotted. In each case, the fur acts as camouflage, an asset to a hunting species.

Solitary and secretive, cats usually live in inaccessible rocky terrain or dense cover. Relying on stealth to catch prey, the felids stalk, then spring upon any prey they can overpower and kill—mammals, birds, fish, even reptiles. Seldom do they wait in ambush for prey to come to them. They usually inflict killing wounds with bites to the neck or throat of larger prey; smaller prey are crushed in the jaws.

Of all carnivores, cats have the largest and perhaps sharpest eyes, directed forward in the skull, befitting a hunter. Cats generally are nocturnal, and their night vision is well developed. The senses of hearing and smell are probably less acute than in canids.

The cats shun human contact, and, though some attacks have been documented, people have little to fear. Research has shown that cat predation on deer and elk helps to keep these herds within the limits of their food resources. Such new knowledge and an increased appreciation for all wildlife give hope that an enlightened human society will provide for the continued existence of the spectacular animals called wild cats.

Mountain Lion (Felis concolor)

Mountain lions, or cougars, do not stalk about "screaming." If they did, they would scare prey animals away and go hungry. Nor does the cougar roar. It makes many sounds similar to those of house cats, but louder. A set of whistlelike sounds—studied in captive cougars—may be used by wild pairs or in families to call and warn each other.

Also called puma, panther, or catamount (cat of the mountains), the mountain lion is the largest of the North American purring cats. Adult males weigh 148 to 227 pounds (67 to 103 kg) and measure 5.6 to 9 feet (171 to 274 cm) from nose to tail tip.

These cats are strict loners. Adult males and females show social tolerance only during the two-week breeding period, and females and

young during the long juvenile dependency period. Breeding is not confined to any one season, but in the northern parts of their range cougars breed mostly in winter and early spring. A pair will remain together for the two weeks, perhaps longer. Then they part; the male plays no further role in the family.

One to six spotted kittens are born after a 90-day gestation period. The spots will give way to the uniform coat of russet or tan responsible for the species name, *concolor,* uniform color. Kittens are born in a cave or other sheltered place, as under a rock ledge or windfall. Helpless at first, they grow rapidly. The mother brings meat to them in addition to providing milk. After about two months they leave the home den and live in temporary dens and caves while the mother hunts.

The young lions possess certain inherent abilities as predators. But they also need to learn, under their mother's tutelage, techniques for killing large prey such as elk. The neck muscles of an elk are too thick for a cougar's bite to be fatal. By

using their forelimbs to twist and snap the elk's neck, cougars in the central and northern Rocky Mountains sometimes kill elk five or six times a cat's weight.

Its training complete at about 18 to 22 months of age, each young lion soon goes its own way.

The cougar once ranged from coast to coast and from central British Columbia to South America. It now occurs in significant numbers only in the areas shown on the map, plus a small, threatened group in Florida.

Biting venison from a kill, a lion uses carnassials, special cheek teeth that scissor the meat.

OVERLEAF: *A blur of speed, a cougar streaks from pursuers. The shy, secretive cats lead solitary lives.*

Ocelot (Felis pardalis)

Garbed to blend into their dappled sun-and-shadow world, ocelots wear beautifully marked coats. And the coats made them marked animals. They and other spotted southern cats were heavily hunted. But the Endangered Species Act made importation of their furs illegal in the United States. Some Latin American countries also officially protect these rare cats.

At home in forest and brushland, ocelots usually rest by day hidden in foliage and at night hunt small and medium-size prey: rabbits, birds, monkeys, pacas, agoutis, iguanas, fish, and frogs. A male and female sometimes roam together.

An adept climber, the ocelot may tree when pursued, though, wrote Ernest Thompson Seton, "He can run like a fox, can blind-hop, back-track, and double-cross his trail." Mexicans who hunted ocelots when they were more abundant not only took the pelts but also consumed the meat and blood. Traditionally, such fare had extraordinary power to abet strength and health.

In cooler parts of their range ocelots tend to bear their young in spring; in tropical areas births may occur in other seasons. A cave or hollow tree may serve as a den.

Information on the elusive species is limited. An average litter is probably two or three.

An ocelot is 36 to 54 inches (92 to 137 cm) long, including a tail of 11 to 16 inches (27 to 40 cm).

Margay *(Felis wiedii)*

Almost a copycat, the margay looks like a small-scale ocelot. Its total length is 32 to 51 inches (81 to 130 cm), including a tail of 13 to 20 inches (33 to 51 cm).

Nearly all that is known about this eye-appealing felid has been gleaned from captive animals. A superb aerialist, the margay likes to climb up to and leap from high perches even when there is no prey for it to pounce upon.

Transferred to a forest setting in Mexico or Central America, such gymnastics would suggest a margay diet of rodents, rabbits, and birds, for its hunting operation can shift effortlessly from tree limb to the ground. The margay's range limit is tenuously put northward into Texas on the basis of a single animal found at Eagle Pass on the Rio Grande in 1852. But the supersecretive habits of this very rare cat may, in part, account for the scarcity of sightings. It is presumed to be nocturnal. By day, the pupils of its large dark eyes, like those of most other small cats, narrow to vertical slits.

Among other common names, the little "tiger cat" is called *chulul* by Maya of Yucatán, and *pichigueta* in Chiapas, Mexico.

Jaguarundi (*Felis yagouaroundi*)

Hued like the desert dusk, the elongated, low-slung jaguarundi can stalk unseen in the half-light. Twilight and dusk are its most successful hunting times. This small-headed southern felid in contour resembles the weasels about as much as it does fellow cats. Tail down, it moves sinuously through the barbed brush called chaparral with scarcely a ripple of leaf or twig to betray its presence. Eventually its body tenses. One

pounce, and a bird in the bush is a bird consumed.

Though an agile climber, this species spends less time in trees than the ocelot. Rodents, other small mammals, and fish from the streams vary the jaguarundi's diet. Preferring to work the ground, it needs no leaf or limb pattern on its pelage. The plain coat may be either a deep shade of gray or russet. Animals of different colors interbreed. Young of both colors may appear in a litter.

The jaguarundi is one of the least known cats on the continent, its life history and population not yet well documented. And now may be too late. Already a rare animal, it becomes even rarer as its habitat—wild thickets and dense lowland

forest—is sheared and put to ranching and farm use.

Mating time for this "otter cat" seems to vary with the locality. The litter of two or three kittens is born after a gestation period known from captive animals to be 72 to 75 days. As in most other cats, kits in the wild probably are cared for solely by their mother.

Full grown, the jaguarundi stands up to 14 inches (36 cm) at the shoulder. Its tail accounts for nearly half its length of 35 to 55 inches (89 to 140 cm). A large individual may weigh as much as 20 pounds (9 kg).

Jaguar *(Panthera onca)*

A favorite hangout of the jaguar is out on a limb dangling over a stream in a sultry southern forest. *P. onca* is very fond of fish. Indians in Brazil believe it flicks the surface of the water with the tip of its tail to lure fish within reach of its claws. It swims expertly, sometimes pursuing the South American caiman in tropical streams. Deer, along with capybaras, tapirs, peccaries, and other small mammals, probably are more frequent prey. In ranch country it may also eat livestock. The jaguar

stalks its prey or ambushes it from a hiding place.

Largest of all the New World cats, the jaguar is heavier than its Old World relative, the leopard, and has a more massive chest. Adults vary from 5 to 8 feet (152 to 244 cm) in total length and 100 to 250 pounds (45 to 113 kg) in weight. The jaguar's short, stiff coat is golden or cinnamon buff, spotted with black rosettes. Melanistic individuals with dark brown or black coats are not uncommon.

The jaguar, known as *el tigre* in Spanish, has a coughing roar of five or six loud guttural notes.

The gestation period is about 100 days. Two to four young are born in a cave or other sheltered spot.

Jaguars once ranged southern California and eastern Texas, and north as far as the Grand Canyon. The last one recorded in California was killed at Palm Springs about 1860. Any resident populations in the United States were eliminated by the early 1900s. Individuals, however, continued to wander up from Mexico. The most recent records are from southern Texas in

1948 and southeastern Arizona in 1949, 1971, and 1986.

In Mexico the jaguar lives in low coastal forests, north along the Gulf Coast to the mouth of the Rio Grande, and on the Pacific side to the Sonoran foothills of the Sierra Madre. From Mexico it ranges southward to Argentina.

This big cat has been relentlessly hunted as a livestock killer and for its valuable coat. Its numbers have been greatly reduced in Mexico, Central America, and southern South America, though it is still common in parts of the Amazon and Orinoco basins.

Lynx *(Felis lynx)*

Feast tonight, fast tomorrow. So goes the hunting pattern of the Canada lynx. Stealthily creeping on big cat feet, spreading its toes where the snow lies soft, the lynx must stalk undetected to within a few bounds of the snowshoe hare or the quarry may escape. Hunter and hunted are well matched. Both are creatures of boreal forest, and so closely is lynx economy linked to the hare that populations of prey and predator peak and crash—almost in unison—once a decade. When a hare depression hits an area, lynxes move en masse, some of them even entering large cities. Though breeding takes place as usual, in late winter, and kittens—one to four—are born 60 to 65 days after mating, few kittens survive in a season of famine.

When times improve, lynx populations begin to rebuild. Born looking and sounding much like the litter of a house cat—but larger and louder—lynx kittens are fed, protected, and taught feline survival ways by the female with no aid from the wild tom. In captivity, however, males have shown paternal solicitude, playing with and grooming kittens.

An old, well-fed male of this north country species may weigh up to 40 pounds (18 kg). To a head and body length of 32 to 36 inches (81 to 91 cm) a stubby tail adds a mere 4 inches (10.2 cm). Females are slightly smaller than males. Larger ear tufts, longer sideburns, and a black-tipped tail distinguish the lynx from its close kin, the bobcat. The long, soft, grayish buff fur of the lynx is lightly mottled with brown. This luxurious fur has made the lynx a target of trappers for 300 years.

Bobcat (Felis rufus)

The stub that earns *F. rufus* its common name is not much. It is too short to be of real use in balancing its owner—which is 28 to 49 inches (71 to 124 cm) long. But the bobcat does manage to climb, scramble, and pounce without rudder-assist. Why a tail at all? Picture a female bobcat out hunting, her three-month-old kittens tagging along in the dense underbrush. The mother holds her tail curved up. Its tip is black only on the upper side. From behind, the white underside is clearly visible, serving as a "follow me" signal to the kittens.

A young bobcat at about seven months old advances a stage in its apprenticeship. It begins to spend time hunting alone on its mother's home range, then returning, perhaps because the hunt has not gone well and it wants the consolation of a full stomach before soloing again. The young become independent at 9 to 12 months and then leave the area where they were reared to find their own territories.

Adult bobcats lead solitary lives except at breeding time, usually

late winter to spring. A male's home range may overlap those of a few females, but they normally avoid encounters by using different parts of the area. During short periods of severe winter weather, neighboring adults may waive the rules and share the same rock-pile shelter, neither threatening nor socializing with one another.

The bobcat, more adaptable to disturbed habitats, has the widest distribution of all Nearctic cats. In recent years it was heavily trapped, and its numbers plummeted. New regulatory controls have helped to stabilize its population.

The high, shady limb of a live oak gives a captive bobcat a vantage point and relief from the heat of day. Reclusive by nature, a bobcat at bay may hiss, spit, and snarl—and stand off many times its weight in dogs or barehanded humans. The mother and kitten opposite put to use the superb sensory equipment of nighttime hunters: keen eyesight, sensitive whiskers, and tufts of hair on the ears that act as antennae to aid hearing.

Though well equipped to find food, cats and other carnivores rarely kill above their needs. Some can live in relative harmony close to people. For others, wilderness areas afford more congenial habitats.

Seals, Sea Lions & Walruses

ORDER Pinnipedia

"NORTHERN FUR SEAL STUDIES" was the subject of my talk to a businessmen's luncheon group. When I had finished and asked for questions, a hand went up in the middle of the room: "Could you explain by what means seals are able to breathe water?" This question so astonished me that from that time on I have started my talks on seals by saying: "Seals are air-breathing, warm-blooded, milk-producing, hair-covered aquatic descendants of land mammals that returned to the sea some 25 million years ago."

The order Pinnipedia (meaning "fin-footed") contains three modern families, which are thought to have originated from terrestrial carnivores. The Otariidae, or "eared" seals, and the Phocidae, or "earless" seals (also called true seals), were once thought to have evolved from different carnivore families. But new research suggests that the entire group sprang from a single ancestral population. In particular, the oldest known fossil seal, discovered recently in California, shows many primitive features that would be expected in the common progenitor of all pinnepeds. The eared seals seem to have developed first, then the earless seals, with the walruses probably being an early offshoot of the latter.

Because the marine environment differs drastically from the terrestrial one, all marine mammals have evolved special adaptations. Powerfully equipped with webbed, five-toed hind limbs and streamlined foreflippers, pinnipeds dive expertly. But earless seals, which are more adapted to aquatic life, go deeper and stay down longer than eared seals. Instruments recorded a Weddell seal descending 1,800 feet (548 m) and remaining submerged for almost an hour. By contrast, a California sea lion's deepest dive fell 1,000 feet (304 m) short of that; and a northern fur seal's longest dive lasted less than six minutes.

Physiological adaptations slow down heartbeat and metabolism and conserve oxygen stored in the blood while diving. To maintain relatively high body temperature in often frigid environments, pinnipeds have developed blubber—fatty tissue under the skin that encases almost the entire body.

Fur seals have such a dense coat of fine fur that water never penetrates to the skin. Fur soiled by oil, however, loses its water-repellent characteristics, and I suspect that seals affected by an oil spill become chilled and die. Even on land, pinnipeds often are exposed to a chilly environment. I found such a place on a springtime visit to San Miguel Island, off California.

The wind, thick with fog, swirled up over a low bluff that bordered a hundred yards (91 m) of sand beach, and the air was heavy with sounds and smells. Barely visible through the mist, a colony of California sea lions was beginning the spring pupping season. Bleating pups, hungry to nurse, searched for mothers. Sleek and dripping, females returning from the sea called repeatedly to their young. Above this bedlam adult males barked as they guarded bits of territory. In the foreground, between me and the sea lion colony, northern elephant seals crowded the beach—seals and sea lions sharing different parts of the same rookery. The elephant seals were mostly weaned pups that had been born in midwinter.

During my visits I have observed differences between the phocid elephant seals and the otariid sea lions on the San Miguel beach. A bull elephant seal, emerging from the water, lay prone. Then he hitched his 5,000-pound (2,272-kg) body up the beach. His great hulk seemed to flow over the sand on rhythmic waves of blubber, with only a minimum of help from the foreflippers. Elephant seals cannot rotate their hind flippers to move forward. But

Flipping sand over his back, a young male elephant seal protects himself from rays of the sun. As a species, the northern elephant seal gets another kind of protection: laws that prohibit hunting.

sea lions can. Progressing on hind legs that extend only from the ankles down, sea lions waddle with a rolling gait, as clumsy as children in a sack race.

Although they exhibit obvious differences, the phocid and otariid seals have much in common. They all eat other marine animals, sharing the characteristic of meat eating with carnivores descended from related ancestors. Walruses glide along the sea bottom sucking up clams. Interestingly, they discard the shells, and only the undamaged clam meats are found in their stomachs. Rinsed in seawater, the meats are a prized delicacy to Eskimos.

Fur seals, sea lions, and harbor seals often come into conflict with fishermen south of the Bering Sea ice pack. Each spring, Steller's sea lions gather by the thousand on the Aleutian Islands adjacent to the waters of Unimak Pass, where salmon migrate toward spawning streams that empty into Bristol Bay. Fishermen assume the sea lions are there for the salmon, but the evidence is lacking. Sea

lion stomachs I have examined contained octopus and small bottom fish.

Many pinniped species are innately tame. This is surely the result of the seals' evolution in an environment free of dangerous predators, at least when on land. Such species as the Alaska fur seal and the monk seals, which inhabit remote oceanic islands, seem unable to overcome their lack of fear in the interest of self-preservation.

Every summer, Alaska fur seals are driven from their hauling-out grounds to killing fields. Certain individuals that, for one reason or another (such as a scarred pelt), are allowed to escape return day after day, even year after year. They make the trip from the hauling-out ground to the killing field repeatedly, seemingly unable to learn they should flee.

Monk seals appear to be extreme in their inability to withstand human intrusion. The Caribbean species was the first large mammal discovered by Columbus on his second voyage in 1493. As Western people occupied the

islands of the Caribbean, the seals gradually disappeared. Having been isolated for millions of years on tropical or subtropical islets, they had not developed the ability to flee and were easily clubbed by anyone who landed on their basking and pupping beaches. As a result, the Caribbean monk seal may be extinct. The last reliable sighting was in 1952.

In 1973 the U. S. Fish and Wildlife Service sent me on an extensive aerial survey of all areas where this seal had been recorded in the last century. The most notable observation we made was that on every island we surveyed we saw signs of human intrusion.

Only two other species of monk seals exist today, the Hawaiian and the Mediterranean. Whether these two relict species can long survive is difficult to predict. Although they appear to be naturally tame, they seem incapable of tolerating human presence when mothers are nursing pups. For example, Hawaii's westernmost island, Kure, was not

A burly northern fur seal bull mates with a sleek female hardly a quarter his size on a beach in the Pribilof Islands. When cows come ashore in the spring, the breeding male gathers a harem on his section of defended beach.

occupied by people until 1960, when a U. S. Coast Guard station was established there. Counts in 1957 and 1958 had indicated that the population of monk seals on Kure was at least 150 animals. Twenty years later only 50 seals remained, all but two of them adults. It was observed that soon after weaning—by abandonment, as is normal—many pups disappeared. To understand why, remember that before people came to Kure most baby seals were born on high, shrub-bordered beaches surrounded by shallow water. After humans occupied the island, however, monk seal mothers moved to isolated sand spits to bear their young. These shifting, storm-swept islets were near deep water prowled by sharks that doubtless found the pups easy prey. As a result, the Kure population may be doomed to the Caribbean monk seal's fate.

Its near relative, the elephant seal, faced extinction early this century. Sealers had systematically slaughtered entire colonies on islands off the California coast. Fewer than 100 sea elephants survived, the nucleus of today's teeming thousands on Guadalupe Island. At Año Nuevo Reserve, on the California mainland just north of Santa Cruz, visitors may approach to within 25 feet (7.5 m) of breeding and basking elephant seals. Their ability to tolerate people has apparently enabled them to regain high population levels.

Today many colonies of seals, protected by laws against exploitation, are returning to ancestral breeding grounds. In some areas controlled harvesting of certain pinnipeds— the northern fur seal, for example—may prove beneficial to the survival of a species.

But in the Bering Sea a sinister new danger looms—fishermen's discarded nets. Made of long-lasting synthetics, they float like kelp, tempting curious fur seals. Many become ensnared and starve to death.

Pinnipeds are highly adapted denizens of the harshest environments inhabited by mammals. One of the most unusual adaptations is the ability to fast for prolonged periods. When the mother Hawaiian monk seal comes ashore to bear her pup, she is enormously fat. In her three-inch (7.5-cm) blubber layer there is enough nutritive value to sustain both herself and her pup for six weeks. During that time she will not leave her pup to find food for herself. Her pup will quadruple its 35-pound (16-kg) birth weight by the time she abruptly weans it. She is gaunt, having lost two pounds for each one gained by her pup.

Among eared seals, males fast during the breeding season. To protect his harem from other males, a bull fur seal must remain on his territory. One that I marked with yellow paint held vigil for 64 days without going to sea to find food for himself.

Pinnipeds are marvelously resilient. When a conservation consciousness began to flourish in the early 20th century, the animals responded. By the mid-1950s most North American species were on the increase. Some, such as the California sea lion and the northern elephant seal, are now at or approaching the maximum populations their habitats can support. Sadly, others, including the northern fur seal, the northern sea lion, the Pacific walrus, and the harbor seal in Alaska, are declining because of new problems— human overharvesting of their prey, entanglement in fishing nets and debris, and deliberate killing by people.

Many wildlife protectionists urge us not to exercise population management in any exploitive sense, but to extend complete protection to all pinniped species. On the other hand, a growing human appetite for fishery products—in areas where seals compete with human needs—leads others to urge control of pinnipeds. These people contend that where large pinniped populations interfere with human efforts to obtain maximum yields from ocean fisheries, the seal numbers should be kept within certain limits.

Regardless of what policies are undertaken, it appears certain that the vast majority of pinniped populations are in no immediate danger of excessive exploitation. But environmental pollution by oil, radioactive wastes, and, indeed, discarded fish nets pose future threats to pinnipeds. KARL W. KENYON

An adolescent Hawaiian monk seal swims between submerged coral heads in the Leeward Islands, which have long been protected as part of the Hawaiian Islands National Wildlife Refuge. Found only in the Leewards, this monk seal was first studied extensively in 1957. Surveys over the years have produced varying estimates of 500 to 1,500 seals, but there is now general agreement that this endangered species is declining rapidly. Reasons include loss of prey, drowning in fish nets, human disturbance of beaches, and pollution.

Family Otariidae

Seal watchers see members of this family from the Alaskan Arctic to Baja California. A pointed snout breaks the surface just offshore. Or a pod of seals basks in the sun and surf (right) of Oregon's rocky coast. They belong to the family of eared seals: 14 species, 4 of which can be seen in the coastal waters of North America. Other species breed on many sub-Antarctic islands in the South Atlantic, South Pacific, and Indian Oceans.

The eared seals are divided into two subfamilies: the Arctocephalinae, or fur seals, with nine species; and the Otariinae, or sea lions, with five species. Fur seals have a more pointed snout than sea lions have. They also have a thick layer of underfur beneath a sparse outer coat of stiff guard hairs. Sea lions have no undercoat.

All members of the family have small external ears and closely set teeth that are sharp and conical—ideal for seizing fish, squid, octopus, and other prey. Their hind limbs can be rotated forward for getting about on land. The fore and hind flippers have bare black palms and cartilaginous extensions beyond the tips of the digits. The fingers of the oarlike foreflippers decrease in size from the first to the fifth; all fingers have rudimentary nails.

Many members of this family—and of the other pinniped families as well—exhibit the phenomenon of delayed implantation. After the female has mated, her fertilized egg goes through preliminary stages of development but then ceases all growth for some weeks or months before becoming attached to the uterine wall and resuming development.

Northern Fur Seal (*Callorhinus ursinus*)

For nearly 200 years the northern fur seal has been exploited for its thick and velvety undercoat—some 300,000 fine hairs per square inch (6.5 sq cm). Recent declines have led to a halt in commercial killing.

During winter, fur seals range through open seas from Japan to the Bering Sea, southward as far as Baja California. With spring's coming, however, they begin to congregate on their far northern breeding grounds. Some 350,000 of them travel to traditional rookeries on the Commander, Kuril, and Robben Islands, all owned by Russia. The greatest population—some 870,000—head for the Pribilof Islands' age-old breeding beaches.

First to arrive in May are the big, breeding males, at least 10 years old and ranging from 6 to 7 feet (1.8 to 2.1 m) in length and weighing about 600 pounds (272 kg). Each bull stakes out a breeding

territory (as the male above has done) and defends it against all rivals. Younger males, unable to compete with their elders, haul out on separate bachelor beaches.

The much smaller females, up to 5 feet (1.5 m) in length and weighing 130 pounds (59 kg), reach the Pribilofs by late June. Within two days of her arrival, each cow gives birth to a black-furred pup. Several days later, the female mates with a bull, whose harem may number 40 or 50.

The mother seal alternates periods of nursing her pup with week-long hunting trips at sea. On her return, she unerringly locates her own pup among thousands of others by its distinctive smell and call. The youngster may double its weight by fall, when its mother leaves the pup to fend for itself.

When the United States acquired Alaska and the Pribilofs from Russia in 1867, the seal colonies were large and thriving. But by 1910 hunters had reduced the herds from some 2.5 million to about 200,000. The Pribilof herd largely recovered by the 1950s but subsequently declined in association with overexploitation, accidental killing by fisheries, and loss of prey. Currently, only about 2,000 may be taken annually for native subsistence.

Guadalupe Fur Seal (*Arctocephalus townsendi*)

This is today a threatened species. Once abundant along the California coast, from the Farallon and Channel Islands to Mexico's Guadalupe Island, this seal was almost wiped out by hunters in the early 19th century. But a remnant survived until 1894 when the species virtually disappeared. Many zoologists considered it extinct.

In 1928, however, two fishermen discovered about 60 of the seals on Guadalupe. They captured two and sold them to the San Diego Zoo.

In a dispute over payment, one fisherman reputedly threatened to exterminate all the rest. He may have almost succeeded, for there was no record of any Guadalupe fur seal for the next 21 years. Once again it was considered extinct.

Then, in 1949, one lone male was sighted on San Nicolas Island off southern California, and five years later a breeding colony of 14 was discovered on the seal's namesake island. Under protection, the group has increased to about 5,000. The main threat today— except for such natural enemies as sharks and killer whales—seems to

be disturbances from tour boats that visit the islands regularly during the breeding season.

One of two species of fur seal in North American waters, the Guadalupe differs from the northern fur seal by having a longer, more pointed snout. Its fur extends beyond the wrist and into the foreflipper's upper surface.

The bull measures about 6 feet (1.8 m) long and weighs 300 pounds (136 kg). Females are considerably larger than northern fur seal cows.

Northern Sea Lion (*Eumetopias jubata*)

You can usually see them on Seal Rocks, near San Francisco's Golden Gate Bridge, and around the Sea Lion Caves on the Oregon coast. They are northern sea lions—also known as Steller's sea lions. That name recalls the man who first studied and described them. Georg Wilhelm Steller was the naturalist with a Russian expedition, led by Vitus Bering, that explored the approaches to Alaska in 1741.

Largest of all the eared seals, northern sea lion bulls range to 13 feet (4 m) in length and may weigh as much as 2,400 pounds (1,089 kg). The females are about 7 feet (2.1 m) long and weigh 700 pounds (318 kg). These sea lions have a coat of short, coarse hair with almost no underfur. Deep-voiced, the big bulls bellow with a throaty roar. Pups bleat like lambs.

An inhabitant of coastal waters and offshore islands, the northern sea lion ranges through the North Pacific from the coasts of Japan and Kamchatka to the islands of the Bering Sea and coastal Alaska, and southward to California's Channel Islands. The world population has declined precipitously, from as many as 300,000 in the 1950s to fewer than 50,000 today. Likely factors include deliberate killing by people, accidental catching in nets and debris, and human overexploitation of prey fish.

Bulls don't eat in the breeding season. They establish territories in early May and collect harems of 10 to 30 cows, guarding them (as the bull does above) until the breeding season ends. Bulls then usually leave to travel northward.

The 40-pound (18-kg) pup is born in late May or early June, and the females mate within a few days after giving birth. Sometimes the youngster nurses for nearly a year.

Northern sea lions eat octopus, squid, crab, and a great variety of fish, diving as deep as 600 feet (182 m) after prey. Fishermen dislike them—and California sea lions—because they sometimes eat commercial fish and damage nets and other gear. For many years Canada worked to control the numbers of northern sea lions in its waters. In a single year hunters in British Columbia killed about 8,000, cutting the provincial population from 12,000 to about 4,000.

California Sea Lion (*Zalophus californianus*)

Adaptable and highly intelligent, this is the familiar performing seal that entertains at circuses and zoos the world over.

Smaller than its northern kin, the adult male measures about 8 feet (2.4 m) in length and weighs 600 pounds (272 kg). A bony crest crowns the male. (The one above barks—never a roar—in Monterey Bay, California, a favorite hauling ground and tourist-boat site.) A female seldom measures more than 6 feet (1.8 m) and weighs a third as much as her mate.

The California sea lion inhabits coastal waters from the Farallons near San Francisco to the Tres Marias Islands off Mexico. The population is about 200,000, divided equally between Mexico and the United States.

A swift and graceful swimmer, this sea lion, like all the eared seals, propels itself with its broad front flippers, and steers with its rear flippers. It is a gregarious and playful animal that sometimes rolls over and over, then pops out of the water like a cork and onto a rocky shore.

Bulls establish territories in the summer breeding season, but females move freely from one area to another. A 12- to 14-pound (5.4- to 6.4-kg) pup is usually born in June. After the breeding season, adult and subadult males often move northward, as far as British Columbia.

A young California sea lion shows 10-mph (16-km/ph) form swimming in the Gulf of California.

Family Odobenidae

The walrus is the only pinniped whose upper canine teeth have evolved into long, downward-thrusting tusks. A moustache of about 400 stiff but sensitive bristles adorns its broad muzzle. The body is thick and heavy— a large adult male weighed 3,432 pounds (1,557 kg)—and the wrinkled, almost naked skin of adult males is marked by many lumps and tubercles. Females and young are covered by short, rust-brown hair. A fold of skin encloses the tail.

Closely allied to the earless seals, the walrus stemmed from ancestral members of that group some 20 million years ago. Like earless seals, it has no external ear cartilage; each ear opening is protected by a flap of skin. Yet, like eared seals, it can rotate its hind limbs forward, and so it can walk on land.

Walruses mostly eat clams gathered from the seafloor. A walrus sometimes dives 300 feet (91 m) when feeding, and it may remain below for up to 10 minutes. A common belief is that a walrus uses its tusks to pry clams loose from the bottom. Zoologists say the muzzle and bristles alone root up the food. Then lips and tongue suck the soft flesh from the shells.

Pacific Walrus (*Odobenus rosmarus divergens*)

After spending the summer months feeding in the Chukchi Sea, herds of Pacific walruses move southward in the fall ahead of the ice pack, passing through the Bering Strait and spending the winter in the Bering Sea. When spring comes, the movement is reversed. Walruses breed on the ice floes in February and March, and most males bear scars of battles for mates. (The bull above rests between rounds.)

Groups of adult bulls—each may weigh about 3,000 pounds (1,361 kg)—are usually the first to head through the strait. Next come adult females with their newborn calves and immature young.

Walrus cows bear only one calf every other year. About 4 feet (1.2 m) long and weighing from 85 to 160 pounds (39 to 73 kg) at birth, the Pacific walrus young nurses for nearly two years. If a calf gets tired while swimming beside its mother, it may hitch a ride on her neck or back. When she dives to the bottom for a meal of shellfish, she sometimes grasps the calf in her flippers and carries it along.

A few walruses become "rogues," eating carrion, or attacking and killing seals and other prey for food. One rogue walrus was seen feeding on a freshly killed narwhal some 14 feet (4 m) long.

Siberian and Alaskan natives have traditionally used walrus hides for their boats, walrus flesh for both human and dog food, blubber for oil, sinews for cordage, tusks for tools and ivory carvings. The illegal ivory market, thwarted by intensified protection of African elephants, may be turning to the walrus and may have contributed to recent declines of the species.

The ATLANTIC WALRUS (*Odobenus rosmarus rosmarus*), slightly smaller than its Pacific cousin, is found from the Kara Sea westward to the Canadian Arctic, and southward to Labrador and Hudson Bay. Once it ranged as far south as the Gulf of St. Lawrence, occasionally to Cape Cod.

Wrinkled hide up to two inches (5 cm) thick serves as effective armor when walruses spar (above). Tusks sometimes exceed two feet (60 cm) in length. They are used as weapons against killer whales and polar bears. Hooked into the edge of an ice cake, tusks help a walrus lever its heavy body out of the water. The animal's scientific name means "tooth-walking sea-horse."

A walrus often uses its tusks to prod a neighbor, signaling it to move over and make room when the animals huddle close on the beach (opposite). If danger threatens, an alert animal bellows to rouse the sleeping herd.

Walruses lying together share warmth. After a prolonged dive in frigid waters, the skin becomes pale (right), for the animal's blood has been concentrated deep within its body. After lying in the Arctic sunshine for awhile, the walrus regains its color.

298

Family Phocidae

Sleek in the sea, ungainly on land, members of this family live all over the world, usually in coastal waters. Five species inhabit Antarctic or far southern seas and nine live in Arctic or northern temperate waters. Monk seals range through warm seas. And two other species—the Baikal and Caspian seals—live in inland waters that are, respectively, fresh and salt. All swim sinuously, moving a streamlined body in somewhat the same way a fish propels itself through the water.

On land, all members of the family move laboriously, with a humping, caterpillarlike locomotion, their hind flippers dragging. Unlike walruses and eared seals, phocids cannot rotate their hind flippers forward. The furred foreflippers help steer and brake. Adult coats are of short, stiff hair with little undercoat. Young of many species are born with white, woolly coats.

These seals range in size from the giant southern elephant seal, which can grow to 20 feet (6 m) long, to the little ringed seal, whose average length is less than 5 feet (1.5 m).

Like walruses, phocids have no external ears. That is why the 19 living species in this family are called the earless seals.

Harbor Seal (*Phoca vitulina*)

A smooth round head with dark, bulging eyes rises from the water, then quickly submerges. Flipping over on its back, the harbor seal swims upside down just beneath the surface, sculling with its hind flippers. Near shore, it turns right side up and humps its way onto the beach.

The harbor seal—also known as the common or hair seal—averages 5 feet (1.5 m) in length and weighs 200 pounds (91 kg) as an adult. Its short-haired coat varies from cream to brownish black in background color, interrupted by countless blotches and spots.

Widest ranging of all seals, the harbor seal inhabits coastal waters throughout much of the Northern Hemisphere. In the Atlantic it is found from the Arctic and coasts of northern Europe southward to France and Georgia. In the North Pacific it ranges from the Bering and Okhotsk Seas southward as far as Japan and Korea in Asiatic waters, and Baja California on the western coast of North America. One landlocked form of the species is found in Canada's Seal Lakes, east of Hudson Bay.

Harbor seals hunt octopus, squid, and other small marine animals, sometimes diving as deep as 300 feet (91 m) and remaining underwater for more than 20 minutes. They also dine on salmon or other commercial fish, and many fishermen do not like them. Canada has had a bounty on them for years, but they are protected in U. S. waters.

Males establish no territories and gather no harems. Mating takes place in the water. Pups, covered with a blue-gray coat, usually are born between March and June on a beach. The pup can swim immediately. Sometimes it must; the next high tide may submerge its birthplace.

Ribbon Seal *(Phoca fasciata)*

This handsome species, also called the banded seal, gets its common name from the distinctive coat of the male (above): a background of deep, chocolate brown, interrupted by broad, creamy white stripes that encircle the neck, front flippers, and the rump. The female is much paler and grayer than her mate. Adults of both sexes average about 5 feet (1.5 m) in length and 200 pounds (91 kg) in weight.

Rarest of all northern seals, the species occurs in two distinct populations: one of about 100,000 in the Bering and Chukchi Seas area; the other—perhaps 130,000 seals—farther westward in the Sea of Okhotsk. There does not seem to be any intermingling of the two.

The Bering–Chukchi population spends the winter and spring along the southern edge of the pack ice, and the white-coated pups are born on sea ice in March or April after a 9½-month gestation period.

Russian traders once used the banded coats as coverings for trunks. Eskimos made clothes bags of the skins. As related by mammalogist E. W. Nelson, "The skin is removed entire and then tanned, the only opening left being a long slit in the abdomen, which is provided with eyelet holes and a lacing string, thus making a convenient water-proof bag. . . ."

Ringed Seal *(Phoca hispida)*

Most abundant of all the arctic seals, the ringed seal numbers in the millions but is essentially solitary. Circumpolar, it ranges ice-bound Arctic coasts. In the Bering and Chukchi Seas, ringed seals move north and south each year with the pack ice. In the Canadian Arctic, they stay in the same general area all year.

The color of their coats varies from gray to blue-black with many creamy, dark-centered rings. Adults of both sexes average only 4.5 feet (1.4 m) in length and weigh about 200 pounds (91 kg). They are the smallest of all pinnipeds.

When the sea freezes over, the seal makes breathing holes in the ice, using teeth and claws to keep them open. A polar bear hunting for food sometimes kills a seal emerging to breathe.

A pregnant seal digs out or adapts a snow den on ice attached to land (below), and there has her pup. She comes and goes unseen through a hole in the floor. But a polar bear may find the den, smash it in, and seize the pup.

The pup is born, white-coated, between mid-March and early May. It will nurse for nearly two months before its mother leaves it.

Harp Seal *(Phoca groenlandica)*

"The ice lover from Greenland," as its scientific name translates, spends the summer near that great northern island, feeding among ice floes of the Greenland Sea and Canada's eastern Arctic.

The population splits in the fall, one part heading for the White Sea, where females bear their pups in late winter, then breed. Another group travels to an area north of Jan Mayen Island and does the same. Members of the third and by far the largest group drift southward with the ice floes until they reach their breeding grounds in the Gulf of St. Lawrence and on "the front"—the ice fields east of Labrador and Newfoundland. There pregnant females haul out on the ice to bear their pups.

Adults of both sexes are about the same size: 6 feet (1.8 m) in length, 300 to 400 pounds (136 to 181 kg) in weight. The male is silvery gray, with a black head and a horseshoe-shaped band from shoulder to flank. This bold marking gives the species its common names: harp or saddleback seal. Markings of the female (above, with newborn pup) are usually paler.

The newborn pup weighs about 26 pounds (12 kg) and has fluffy white fur that gives it the name "whitecoat." It nurses for 10 to 14 days, increasing its weight nearly threefold—mostly blubber—by the time it is weaned and abandoned. The pup lives off its fat and molts its white coat for a smooth coat of gray hairs.

Harp seals, mainly pups, were heavily hunted for their skins. Kills up to half a million a year resulted until public outcry caused American and European bans on imports. Since 1984, hunting has continued at a reduced level, with approximately 60,000 seals taken a year. The overall Canadian population of harp seals is now thought to be stable at around 2 to 3 million.

Gray Seal *(Halichoerus grypus)*

Every summer a hundred or more gray seals appear in Maine waters, hauling out to sun themselves on the approaches to Mount Desert Island and the islands of Penobscot Bay. Some 200 miles (322 km) to the south, a band of 15 or 20 live year-round on shoal islands in Nantucket Sound. There the pups are born in January or February in the ice off Muskeget Island. These are the southernmost breeders of all the gray seals in the western North Atlantic.

Gray seals have three distinct centers of population: the waters of southeastern Canada; northern Europe from the British Isles to the White Sea; and the Baltic Sea.

The total population of about 130,000 may be increasing in many areas. Canada's gray seals, numbering some 30,000, seem to be thriving. Nearly 2,000 pups were counted on Nova Scotia's Sable Island in a recent year.

Males and females differ in size (as the two above show). A big bull may measure 9.8 feet (3 m) and weigh 640 pounds (290 kg). Females usually measure some 7.6 feet (2.3 m) and weigh 550 pounds (249 kg). The bull's coat ranges from dark gray to almost black, with many spots and markings. The female's lighter background has dark spots. A prominent nose gives the species another name: horsehead seal.

Unlike most seals, bulls are territorial, gathering harems of pregnant females. A pup weighs 30 pounds (14 kg) at birth and may triple that weight when it is weaned at about three weeks of age.

Many fishermen look upon the gray seal as a thieving nuisance. For a number of years the Maritime Provinces of Canada offered a bounty on the species, and Nova Scotia conducted an annual "cull" of the seals under the supervision of fisheries officers. Both practices now have been discontinued.

Hooded Seal (*Cystophora cristata*)

A unique bit of anatomy, the male seal's nasal pouch, gives this species its common names: hooded, crested, or bladdernose seal. The pouch, also called a hood, usually hangs limp and wrinkled over his nose. But it can be inflated to twice the size of a soccer ball. Experts debate why the male swells his pouch. It may be excitement. (Or anger, which the male at right may be showing a photographer.) An inflatable nasal membrane—it looks like a bright red balloon—can also be pushed out of a nostril.

An adult male averages 6.5 to 10 feet (2 to 3 m) and weighs 700 to 900 pounds (318 to 408 kg). His hair is bluish or gray, marked with darker spots and blotches. The female (in foreground above, with pup) is smaller and paler than her mate and lacks a nasal hood.

Most of the world's 500,000 hooded seals live in the East Greenland Sea and breed north of Iceland and Jan Mayen Island. Between 50,000 and 75,000 breed in the pack ice off Labrador, Newfoundland, and the Gulf of St. Lawrence.

Hunters kill as many as 15,000 hooded seals each year in Canadian waters during breeding season.

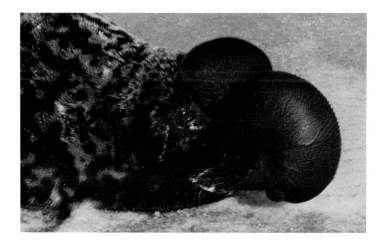

Northern Elephant Seal (*Mirounga angustirostris*)

The elephant seals, largest of all seals, get their common name from the trunklike proboscis dangling from the muzzle of adult males. The snout is an extension of the nasal cavities and, when inflated, curves into the seal's mouth. This acts as a resonating chamber in the throat when the bull snorts or bellows. This elephant seal and its Southern Hemisphere kin, *M. leonina*, are anatomically similar.

The northern species once inhabited breeding rookeries and hauling grounds on offshore islands from Baja California to the Farallons. In the 19th century they were ruthlessly hunted, for a large bull—an easy victim—could yield up to 200 gallons (756 l) of fine oil. The species was almost extinct by 1890; fewer than 100 survived at Guadalupe Island, Mexico. Then, in 1922, the Mexican government began protecting them. They steadily increased, and today they number about 100,000.

Adult bulls come ashore on the breeding beaches in November and December, fighting one another as they establish territories. Karl Kenyon describes an encounter: "An established bull rose up to offer a bellowing challenge to the intruder, and the two aggressive beasts faced each other chest to chest before separating to rest while eyeing one another on either side of an invisible but well-established territorial boundary."

Males may measure 17 feet (5 m) and weigh 5,000 pounds (2,268 kg). Cows range to 11 feet (3.4 m) and 2,000 pounds (907 kg).

Pregnant cows arrive at the hauling-out grounds in December; within a few days each gives birth to a dark-haired pup. (An angry mother guards the one above.) By the time the pup is weaned—at about four weeks—it has tripled or quadrupled its birth weight.

Dueling for dominance, bulls bleed but rarely die. The fight may be a one-minute round, ending when the vanquished gives ground.

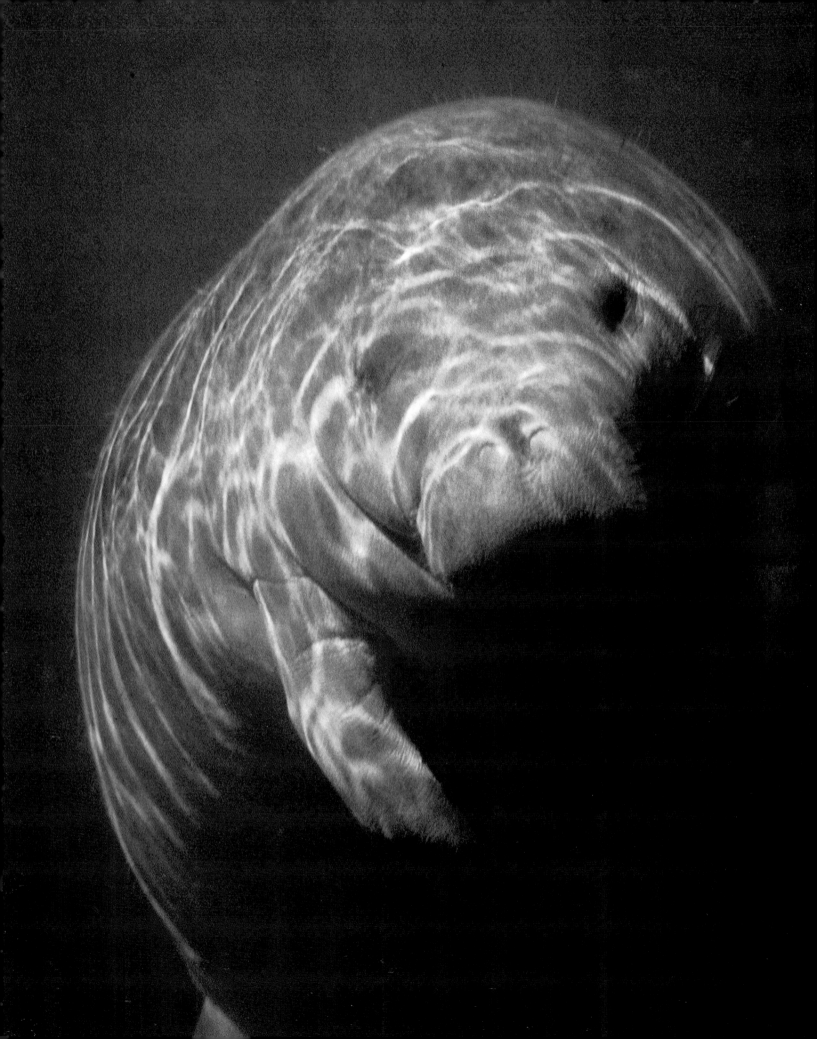

The Manatee

ORDER Sirenia

HOMELY AND UNGAINLY, without the grace of dolphins or the grandeur of whales, the sea cow seems an unlikely inspiration for ancient tales of alluring mermaids. But presumably the animal did inspire such stories, and it remained for explorers of the New World to explode the myth. Grumped one 16th-century Spanish chronicler, "So ugly is [the manatee], that uglier it cannot get."

Today only four sea cow species survive, most of them living in shallow tropical marine waters and the larger tropical rivers of the world. A fifth species, Steller's sea cow, *Hydrodamalis stelleri,* roamed the frigid Bering Sea until fur and seal hunters exterminated it around 1769—fewer than 30 years after it was first reported.

Three of the still surviving species are called manatees. They belong to the family Trichechidae. The fourth is the dugong, of the family Dugongidae. And all are called sirenians—in memory of their mythic beauty.

Only one species, the West Indian manatee, *Trichechus manatus,* frequents the waters of the continental United States. It ranges northward along Florida's Gulf and Atlantic coasts—occasionally as far as the Carolinas—and south to below the Amazon River. (Another species, *T. inunguis,* confines itself to the Amazon River and its tributaries, and an African species, *T. senegalensis,* ranges the coastal waters and larger rivers from Senegal to Angola.)

Sirenians are large, seal-shaped animals with front flippers and a horizontally flattened tail. They are warm-blooded and breathe air. Adults vary in size from about eight to twelve feet (2 to 3.5 m) and weigh from several hundred to more than a thousand pounds (454 kg). (The extinct Steller's sea cow, a giant among sirenians, attained a length of 20 feet [6 m] or more.) The thick, tough skin of sirenians is only sparsely haired, except for numerous stout bristles on the face.

Sea cows probably share a common ancestry with elephants, whose fossil remains date back some 50 million years. Sirenians' molars, like those of elephants, are continuously replaced with new teeth that move forward from the back of the jaw as the old ones wear down and are pushed out.

These animals are unique among mammals in that they are the only herbivores to spend their entire lives in the water. And they are gentle. A sea cow will not fight—even to save its own life. When faced with danger, its only defense is a hasty retreat at speeds of up to 20 miles (32 km) an hour.

Sirenians spend much of their time grazing on submerged beds of sea grass and other aquatic vegetation. Several animals may swim together in a small, loosely associated herd or, in winter, groups of 20 or more may congregate around a warm spring. On such occasions they nuzzle and play and seem to enjoy one another's company.

The dugong generally resembles the manatee in size and shape, except that its tail is divided into flukes, while the manatee's is rounded and paddlelike. Dugongs inhabit Indo-Pacific waters from the Ryukyu Islands south of Japan to northern Australia and along the coasts of Mozambique and Madagascar. Some are found among the islands of the U. S. Pacific Trust Territories.

The fate of sirenians today is precarious throughout their range. In many parts of the world, human activities have menaced the animals' survival. Even in protected areas they are still hunted for food. And their birthrate (usually one calf per cow every two or three years) may not be enough to prevent their eventual elimination from the roster of mammalian life.   HOWARD W. CAMPBELL

Manatee (*Trichecus manatus*)

The manatee still appears now and then in coastal waters from North Carolina to Texas, but its only stronghold in the United States is Florida. Fewer than 2,000 of these placid mammals are left—remnant herds scattered along the state's Gulf and Atlantic shores.

Widely hunted in the past for their oil, hide, and meat, Florida's manatees face dangers of a different kind: collisions with powerboats and destruction of their habitat by urban development.

Small eyes and a droopy muzzle give the manatee a forlorn look. How well can it see and smell? No one knows. But it does hear well.

Manatees often communicate among themselves with a varied repertoire of squeaks, bleeps, and chirps.

Manatees lack leg and hip bones, but their front flippers retain vestigial nails and a handlike skeletal structure, reminders of a time when these animals lived on

land. Nor does the manatee use its flipppers to swim. Its tail propels it, stroking powerfully up and down. The flippers are used mainly for close maneuvers—and to embrace during sexual congress.

Sea cows cannot stand cold water. During Florida's cool months they move southward, seeking the warmth of freshwater springs—and even hot-water

outlets from industrial plants. (The bull calf above, in a tattered coat of algae, shares his warm spot on the Crystal River with a school of gray snappers.)

Sea cows bear their young after a gestation of up to 400 days. Calves average 50 pounds (23 kg) at birth and suckle for two years.

Like a blimp in midair, a manatee hangs suspended in its element. Sea cows surface for air every two to five minutes but can hold their breath up to 20 minutes at rest. Valved nostrils seal out water when they dive. Dense, ivorylike bones enable them to stay submerged with minimal effort.

Hoofed Mammals

ORDER Artiodactyla

In the bone-nourishing velvet of summer, a white-tailed buck grows antlers that will take him through autumn's rut—testing time for him, a time of renewal for his kin in the order Artiodactyla.

AMERICA IS A LAND OF GIANTS. After the dinosaurs died and the mammals arose, great beasts again roamed the land. Travelers across the land bridge that at times connected Alaska with Asia, they came from abroad or originated in this hemisphere and spread to the Old World. A wanderer through the ages could tell us of unbelievable creatures that once lived here, of the great extinctions and the turmoil of invasions as the bridge rose and fell with the pulse of glaciations.

That is certainly true of the order Artiodactyla, the cloven-hoofed mammals—no less so than other orders. Bizarre colossi—multihorned relatives of deer with slingshot projections on their snouts, camels more than 10 feet (3 m) tall, piglike animals as big as cows—ranged the continent. In the fossil record, the artiodactyls are found back to the beginning of the age of mammals. They proliferated into many lineages and showed their greatest abundance in the cooler, drier epochs preceding the ice ages, when grasslands covered much of North America.

The giants are still here. Take almost any large mammal that Eurasia shares with North America and you will find the largest form on New World soil. The Siberian moose is large, but the Alaskan moose is larger. The giant among reindeer, the Osborn caribou, dwells in the Cassiar Mountains of British Columbia. Eurasia has many species of Old World deer; only one penetrated into North America—the elk, or wapiti, the giant of the tribe. So it goes, not only with contemporary animals but also with extinct forms. The largest bear, lion, saber-toothed cat, wolf, bison, goat-antelope, elephant, horse, beaver—and on and on—is found in the fossil record or is still alive in North America.

The appearance of giants on this continent is not difficult to understand. A large animal has proportionately less skin surface in relation to body volume than does a small animal. In cold climates this makes large size an advantage for retaining body heat.

When two members of a species compete for food, the larger one can usually find more to eat. And as a large mammal species disperses and colonizes new regions, the animals' new-found plenty tends to make them grow even bigger. We find, in the course of an evolutionary radiation, that small-bodied forms that develop in the tropics spin off increasingly larger-bodied and more grotesque forms as they invade progressively colder or drier environments.

North America, as an adjunct to the Eurasian landmass with a periodic northerly connection, would receive mainly giants adapted to cold and dryness. Had the Bering land bridge been at a lower latitude, the fauna here would be much more similar to that of Asia than is the case today.

It is also easy to understand why giants have such a high rate of extinction. They generally are ecological specialists, and therefore much more sensitive to ecological changes than smaller, more primitive, broadly adapted species. Such smaller animals have the best record of survival. Fossil beds are filled with the bones of extinct giants.

The severest slash to the American fauna came at the end of the last glaciation. We call it the megafaunal extinction. And we still argue about its cause (page 383). Was it climate? Was it humans?

This background of evolutionary development is necessary for an understanding of the peculiar pattern of occurrence of artiodactyls in North America today. The order is composed of three suborders, all very distantly—if at all—related. They are Suiformes, which includes pigs, peccaries, and hippos; the Tylopoda, containing camels and llamas; and the Ruminantia, encompassing the deer, giraffes, bovines, pronghorns, and mouse deer. Altogether some 220 species occur, with representatives of the order on all continents except Australia and Antarctica.

Ironically, the camel and the horse (order Perissodactyla) originated in the Americas. Very late in their evolutionary history they entered Eurasia and Africa, and then became mysteriously extinct in their homeland. When

When antler growth has ended, the peeling velvet is rubbed off on tree trunks, bushes, or branches. Then a crowned warrior enters the rut, equipped with both a weapon against other males and a signal to females awaiting a mate.

After the rut is over, the antlers drop off. In spring new growth begins (opposite) from bone stumps called pedicels. Antlers differ in structure from horns, which are permanent projections.

Jefferson Davis, as Secretary of War, imported some camels for Army use in the Southwest before the Civil War, stockmen were astonished at the animals' appetite for such local plants as creosote bush and manzanita, which other livestock spurned.

Camels did not survive the prejudice of mule skinners and ranchers who shot them. Horses and burros, however, survived in a feral state and flourished as true natives. Which, of course, they are—natives reintroduced from

Europe. They were living in North America long before such species as the bighorn sheep came here.

Bighorns came from eastern Siberia; only on our continent has this northern, arctic form penetrated deep into the south. Desert bighorns are found from Nevada and California into Mexico. But when a true "desert" sheep—the Barbary sheep—was introduced from Africa into New Mexico, it readily accepted dry wasteland regions, where now the bighorn barely hangs on.

Artiodactyla means "having an even number of toes." And an even number of weight-bearing toes is one of the few things that members of the order have in common. A unique ankle joint and a foot structure based on development of the third and fourth digits (toes) carry the weight of the animal. This evolved as a means of handling the stress generated during running. The harder the ground and the higher the speed and the heavier the weight of the animal, the simpler and more elegant the foot structure became.

The horse's weight is carried by the third digit—a single hoof. In the artiodactyls, two digits do the work. This results in a symmetrical foot that has two main hoofs and often two small side hoofs, or dewclaws.

The megafaunal extinction left North America with only four families of artiodactyls. They are the Cervidae, the deer tribe; the Bovidae, hollow-horned animals that include bighorn sheep, mountain goats, and bison; the Antilocapridae, or pronghorns, which are truly American natives; and the Tayassuidae, or peccaries, small, piglike creatures of American origin.

Except for the peccaries, all belong to the suborder of ruminants. Members of this suborder have a sophisticated digestive system, which enables them to exploit plant cellulose for energy. No mammal has enzymes capable of digesting cellulose; the ruminants culture microorganisms—bacteria and protozoa—to do it for them. These organisms grow within the rumen, an organ where fodder ferments by bacterial action. Fermentation

In the first snowfall of winter young bull elk spar and posture, performing a continual and ritualistic test of strength. Duels for dominance regulate social rank among hoofed mammals.

produces fatty acids and other products that enter the animal's bloodstream directly from the rumen. Then the bacteria and residue from the fodder go into the stomach and gut, where the bacteria are digested to further supply the ruminant's protein, mineral, and vitamin requirements.

Most true ruminants have some kind of hornlike organ sprouting from their heads. The projections may be ossicones—permanent bone cones covered by dense skin and hair, as in giraffes. They may be a tough horn covering a bony core—permanent and increasing by a segment each year, as in the bighorn. Or they may be bone structures grown and shed yearly, as are deer antlers. Horns and antlers serve as weapons of attack

and defense, and for display purposes. They may be luxury organs, in the best sense of the word. The better an animal is at getting the most nutritious food, the more it can spare from pure body needs.

The excess can be put into luxuries, into status—as big horns. Thus horns become symbols of prowess and ability. They say that physiologically this is indeed a very capable animal and a very good forager. Since the male with the biggest horns will breed with the most females, his traits will be passed on to more offspring.

But there's another side to having the biggest horns. My studies with mountain sheep have shown that rams with the largest luxury organs—the most dominant, the most

virile, the most successful individuals—have shorter lives. The same probably holds true among other ungulates—hoofed mammals. Such males exhaust themselves fighting and in rutting activities and mating. The result is that the less active, less prosperous male has the longest life expectancy.

Why do the females of some species have horns? Among ungulates that live together in large herds, females often look like males—so much so that even seasoned biologists may be fooled. One theory is that since the female must compete for forage at times against young males, she assumes their image. Another reason the females may have evolved horns is to discourage courting and harassing advances by young males, which tend to

stay with the females after the rutting season. On the tundra, caribou cows grow antlers equal in size to those of three-year-old bulls. But among some of the more solitary woodland caribou, up to 40 percent of the females do not have antlers. Female elk don't grow antlers, but they do have manes and are nearly as large as the bulls.

If antlers are so important, why are they shed? Bucks busy sparring, fighting, and courting during the rutting season have little time to forage. After the rut they are exhausted, weakened, and vulnerable. Antlers make them stand out from the females; therefore, it's likely that predators would quickly learn that such standouts can be easy prey. Thus bucks have evolved the trait of shedding bone growth in order to look like just one of the crowd. Male mule deer not only slough their antlers but also mimic female actions— even crouching to urinate.

Another strategy for avoiding predators after the rut would be to go into hiding. And that is what the male blacktail and mule deer do. Bull elk do too, for a time. But then they form their own groups, separate from the females. At that time, however, the males need their antlers to maintain their dominance

hierarchy. Consequently, they don't cast off their racks until warm weather.

In the north, bull elk barely have enough time to grow and harden a new set before another rutting season begins. But for elk populations established farther south—the tule elk of California, for example—the seasons have so spread that males begin to rut while their new antlers still are in velvet.

Strategies against predators involve more than just the shedding of antlers. Take the case

of the mule deer and the whitetail. Both may share the same habitat. But the whitetail is a hider; it favors dense brush, and if discovered it erupts from cover with a crash, then dashes off. It runs as fast as possible, trying to get so far ahead of pursuers that its scent may evaporate and dim the trail. Or it will run through water or swampy spots to avoid leaving a trace of scent. It also likes to run where other deer may be lurking—and let someone else get chased for a while. Such habits make for a nervous, fidgety species.

The mule deer pussyfoots around in the brush, coolly trying to stay hidden. If flushed, it "stots"—jumps straight up and possibly changes direction. While the whitetail prefers to rush downhill (letting gravity help add to its speed), the mule deer likes to bound uphill. This lets it gain elevation comparatively easily while the predator is forced to clamber up at far greater cost of energy. It also enables the mule deer to leap over a boulder or bush; the pursuer has to go around.

Bounding also permits the deer, with each jump, to depart in a totally unpredictable direction. But with the predator close at hand, the deer must not jump until the last possible moment. Otherwise, the pursuer can redirect itself. Such escape techniques call for coolness in timing—and make for a calm, collected type. Thus, as one offshoot, tranquil mule deer can be tamed with ease. Jittery whitetails cannot.

In tundra country or on open plains with scant cover, a lone newborn calf can be particularly vulnerable to predators. But if many young are around, their numbers swamp the enemy; a few calves may be taken by predators, but a lot live. So genetic selection among such herding mammals as the caribou has made for females that not only all reach a breeding state at the same time but also give birth almost en masse.

Among species whose young instinctively hide, such as the white-tailed deer, females give birth at different times. This reduces the risk that predators, taking one easy prey, will learn to expect to find others. Among such

"hiders," the birthing span tends to spread. So does the rutting season. Warm climates—with milder conditions for the vulnerable newborn and longer periods of nourishing vegetation—present more opportunities for rutting and giving birth. In such regions, females readily extend the birth season.

Except for pronghorns, single young are the rule among animals that live in herds and run for their lives. The larger the young, the more likely it will be able to outrun predators.

Still shedding its long-haired coat of winter, a mountain goat clambers up a steep slope of its lofty habitat. Hoofs (left) with a hard outer edge for digging in and a tough, rubbery inner pad for traction make the goat at home on cliff edge, crag, or smooth rock. Members of the order Artiodactyla have feet designed so that two toes carry the weight. Adaptation produces variations, such as the caribou's spreading hoof, which is suited for travel over soft ground.

And it will grow out of its dangerous juvenile period faster if it does not have to compete with a sibling for the mother's milk. The baby gnu of Africa is a classic example; five or ten minutes after birth it can keep up with a herd that is running 25 miles (40 km) an hour.

While the ungulate mother bears and cares for her young, the father contributes only his genes. Therefore, this contribution should come from a superior male. To be fit, one must maintain access to resources, particularly when they are scarce. Fighting over resources can be costly in time and energy, to say nothing of the danger from horn or hoof. The cheapest way of maintaining access is to establish dominance over competing individuals. Then the most dominant animal

A caribou calf a few days old trots after its mother across the Arctic tundra. Northern animals that travel in herds have their young—one big baby for each mother—at the same time, virtually on the run. Scattered births or litters would give predators an easier and a steadier diet.

gets the biggest share and the other animals get progressively less.

A male ungulate's ultimate aim, of course, is access to the female. Some ungulates—such as mule deer and moose—are sequentially polygamous: The male stays with one female until she is bred, then he leaves her for another. In contrast, elk bucks herd females together into harems for mating.

The bull elk advertises. He bugles. He tries to outbugle his competitors, and if one advertises nearby, he goes and shuts him up or chases him off. The bigger the elk's body, the more resonance his call has. So the females cluster to the bull with the deeper voice and the one that other bulls can't shut up.

But fighting for dominance among ungulates can cause serious wounds, even death. The sharp horns of mountain goats jab and puncture, and the males have evolved a tough shield—hide an inch (2.5 cm) thick— on their rumps, where the most blows land in their side-to-side, head-to-haunch type of fighting. Even so, deep wounds result.

The elk charges with sharp polished antlers; punctures and broken necks may occur. Bighorns clash head on. They have evolved multiroofed skulls, tough facial skin, and incredibly sturdy neck tendons to withstand tremendous impact. Moose can kick from both ends. A bull moose, using its hind legs, struck a man standing on a corral chute eight feet (2.5 m) above the ground.

Injury to the victor, however, may be as severe as to the vanquished. So real fights are infrequent. Instead, ungulates have adopted more subtle methods that aim at the same establishment of superiority but are not as costly. They are called dominance displays.

Essentially, display occurs when an animal shows off its weapons without making a move to use them. Threats, on the contrary, bring weapons into readiness. Deer that rise on hind legs to flail may threaten by lifting the head or forebody, raising a leg, or stamping the ground. The cow elk that bites will pull back her lips and grind her teeth. Or a dominant bighorn ram threatens an inferior competitor by jerking his horns downward in order to frighten him. But to a serious rival he displays his horns with his head drawn up and back so that his neck muscles bulge. Thus, the opponent may judge his power.

The displays of most large mammals show off the mass of their bodies. To display, these animals turn broadside and erect specialized hair, such as hair of the mane or along the spine. They call attention to distinctive markings and make themselves appear as large and conspicuous as possible.

The red deer, close relative of the elk, enhances the size of his antlers by horning shrubbery and collecting vegetation on his rack, so it looks larger. During the rut, the bull elk urinates on his underbody, then wallows in wet soil. The mud darkens the animal along his entire length, including the mane, making him more impressive.

But it isn't enough just to display. The dominant animal has to have credibility: Now and then he has to back up his threats with force. The ultimate credibility is a test of strength. It pays the dominant animal to reinforce his position periodically—and the subordinate animal to test the dominant's competence from time to time. So sparring, even serious fighting, results.

Interestingly, mule deer form social bonds as a result of sparring matches. A superior may defeat a lesser male in successive bouts and lord it over him afterwards, but if the loser is harassed by other lesser males, then the superior comes to his rescue.

Sparring and fights both are ended by the subordinate's breaking off with some submissive act. Pretending—with exaggerated motions—to graze is an almost universal signal among hoofed animals that says, "Look, I'm really peaceful." Often the submission takes the form of acting like a female.

Thus the effect of weapons, the ability to learn, the preference of individuals to live in a social milieu where roles are understood and played out—all this combines to create a dominance hierarchy we can recognize. They sometimes act like us.   VALERIUS GEIST

Family Tayassuidae

Peccaries aren't pigs, but superficially the animals are look-alikes. The two families began going their separate ways some 40 to 70 million years ago when the pigs (family Suidae) developed in the Eastern Hemisphere and the peccaries mainly in the west. Thus, in North America today both the domestic hog and the wild boar *(Sus scrofa)* are introduced "exotics."

The less obvious differences between pigs and peccaries are considerable. Peccaries have fewer teeth than most pigs (38 to a pig's 44) and have partially fused foot bones, an adaptation for running. Peccaries' shorter, straighter tusks fit so closely that they hone each other to razor sharpness with each snap of the jaws. These spear-edged weapons give peccaries a common name, javelina.

Peccaries are more herbivorous than pigs and have a more complex stomach for digesting coarsely chewed food. One favorite of the collared peccary: cactus, especially the prickly pear, which is eaten spines and all.

The family's three species—placed in two genera—range from the southwestern United States to central South America.

Peccaries have on their rumps glands that give out a fluid with a strong, musky odor. Early observers of the "musk hog" mistook the gland for a navel.

Collared Peccary *(Dicotyles tajacu)*

These peccaries usually live out their lives in social groups of 5 to 15, though some, old and infirm, may die in solitude. There are no harems or bachelor groups. The herd sleeps, forages, and eats together.

But there are squabbles. Adversaries square off, lay back their ears, and clatter their canines at each other. In fights they charge head on, bite, and occasionally lock jaws.

Scent from the rump gland is a cohesive of the herd. The scent marks the home range and helps herd individuals identify each other. Peccaries rub against rocks, stumps, and tree trunks, leaving smears of the oily fluid. They also rub against each other, standing head to rump. This is apparently a form of greeting, not a sexual ritual.

There is no certain breeding season. Young, usually twins, are born about 144 days after mating. When mature, they weigh 30 to 60 pounds (14 to 27 kg).

D. tajacu ranges from southern Arizona, New Mexico, and Texas into northern Argentina.

These twins will follow mother past their first year; as infants, they sheltered beneath her, their purring answered by grunts.

Family Cervidae

Long-legged and graceful, delicate or formidable, North American members of the Cervidae, the deer family, range from tiny Key deer of Florida to huge Alaskan moose. Males of all American species, and caribou females, carry antlers. These grow rapidly in summer—soft, tender bone covered with a thin skin whose fine hairs look like velvet (as the moose, right, is wearing). When growth stops, the skin dries and is rubbed off. Later, a ring of cells breaks down bone at the base; antlers drop—to weather away or be gnawed by mineral-hungry rodents. Antler growth depends a great deal upon the animal's health; the number of antler points does not indicate a deer's age.

Like other artiodactyls, deer have fused metacarpal and metatarsal bones—structures analogous to human palm and instep bones—which form the shock-resistant "cannon bone" of the lower leg.

The deer family originated in the Old World about 40 million years ago. But some species that evolved on this continent, often called the New World deer, have a relatively primitive foot equipped with well-developed dewclaws and elastic hoofs, ideal for soft ground and for climbing. North American species live in forests, upland deserts, swamps, and tundra. Old World members, as well as their North American counterpart, the elk—or wapiti—have hoofs better suited for hard ground.

Elk or Wapiti (*Cervus elaphus*)

The two bull elk walking side by side break into a trot, then a run. Two hundred yards (182 m) across the meadow they wheel and walk back. Several times they pace, sweeping antlers low to the ground at each turn. Glinting antlers and dark body parts catch the rival's eye to show off size and strength. Thus they duel for dominance.

Special brain cells—in animals as well as humans—react to sharp edges moving across the line of vision. Elk antlers present such sharp edges. But the antlers are not merely symbols; they are sturdy enough to absorb a rival's thrust—and straight enough to prevent entanglement. Antlers rarely lock.

Elk are the most highly evolved of the Old World deer. They have adapted to open land, feeding on grasses as well as forest browse. The raised head helps an elk in the open spot a predator. The animal spends about one hour in seven standing or walking; the rest of the time, feeding or resting. So a group needs at least seven members as insurance that on the average one head will be up (as happens in the photo above).

Elk, called wapiti by Shawnee Indians, stand up to 5 feet (152 cm) at the shoulders and weigh as much as 1,100 pounds (500 kg).

The TULE ELK (*C. e. nannodes*), palest, smallest subspecies, weighs 325 to 400 pounds (147 to 181 kg). Its habitat is confined to a small area in east-central California.

Sunrise gilds an elk mother and her suckling. Birth—usually one young—is in spring. By four months a calf is weaned and unspotted.

Mule Deer (*Odocoileus hemionus*)

Bleating like a young deer in distress, a buck trots after a doe. He is trying to get near enough to coax her to urinate. Then he will sniff and curl his lip so that a nasal organ can analyze her urine; the level of hormones will tell whether the doe is ready to breed. But the doe has a genetic mission: to be selected by the fittest. And she skips along, retaining her urine. Now the buck tries another courtship strategy. With a sudden leap, a slap of the ground, and a horrible roar, he frightens her so much that she urinates. From such autumn rituals come in May, June, or July the birth of young, usually twins.

The mule deer gets its name from huge ears two-thirds the length of its head; they aid in detecting danger at long range. Mule deer are also called jumping deer because of their stotting— stiff-legged bounding with all four feet off the ground—when sensing danger. They can stot while standing or in flight.

Mule deer differ in individual body markings and coat colors. Tones range from a dark ash-gray or brown-gray to a light gray or browns and reds. Rump patches, which surround the short, slim, black-tipped tail, range from white to yellowish.

Mule deer forage for a wide array of foods. Twigs, leaves, and rotted or frost-killed plants are consumed in small bites. The deer may climb hard snowbanks to reach aspen catkins high above the ground.

Though its forebears predate the ice ages, the mule deer is a relatively recent and advanced offshoot of the genus. It has adapted to high elevations and semidesert regions, and can be found in broken country, chaparral, brush, and woods. Adult males vary in weight—from 180 to 400 pounds (82 to 181 kg)—and stand about 3.3 feet (101 cm) at the shoulder. Females are smaller.

Columbian Black-tailed Deer *(Odocoileus hemionus columbianus)*

One of eleven subspecies of *O. hemionus,* the Columbian blacktail haunts forested coastal regions from southern British Columbia to central California. Browsing on lush undergrowth, it seldom needs to drink.

This subspecies and the Sitka deer *(O. h. sitkensis)* are often called black-tailed deer, while the others are usually known as mule deer. Sitka deer live in British Columbia, Alaska's panhandle, and offshore islands.

In the mountainous eastern limits of its range, the Columbian blacktail breeds with mule deer. Typically, *O. h. columbianus,* with its distinctive brushlike tail, is darker and smaller than the muley. A blacktail buck stands about 3 feet (91 cm) at the shoulder; weight ranges from 110 to 250 pounds (50 to 113 kg). Does average about 100 pounds (45 kg).

While mule deer, adapting to progressively higher elevations and scattered forage, have extended

☐ Mule Deer
☐ Columbian Black-tailed Deer

their range across western North America, blacktails are found only in the Pacific Northwest.

Less social and more nocturnal than mule deer, and with localized migratory tendencies, blacktails in small bands have been observed defending territories against intruding deer. Thus, by not sharing the available food, they stand a better chance of surviving the winter. In defending an area, blacktails demonstrate individually and collectively the ability to recognize certain landmarks that define territories.

White-tailed Deer (*Odocoileus virginianus*)

Six veteran hunters combed the square mile (2.5 sq km) of forest where 39 deer were fenced. It took the hunters four days to spot one buck. In another test, five men followed a buck wearing a radio collar and orange streamers on his ear tags. They took all day to find him hidden in the underbrush. Thus does the white-tailed deer merit its proverbial reputation as a hider—and survivor. For the animal that supplied buckskin and venison to a frontier nation still exists in great numbers: Urbanized Pennsylvania can sustain a yearly harvest of about 100,000—and some 25,000 more killed by cars.

Except for parts of the Southwest and California, whitetails range over most of the continent south of Hudson Bay. Some 20 subspecies live in the Nearctic. Northerners (such as the whitetail buck and fawns above) look chunkier

Key Deer (*Odocoileus virginianus clavium*)

The young leave a hoofprint the size of a human thumbnail. The adults themselves grow no larger than an Irish setter or German shepherd. These are the Key deer, a diminutive subspecies of white-tailed deer.

Found only on the Florida Keys, they are protected as endangered animals. A few decades ago, their numbers had dwindled to 25 or 50. Now there are about 250, but their habitat is encroached upon by humans, and about one a week is killed on a highway. A natural limitation—availability of fresh water—also checks their population. The Key deer has a reproductive rate lower than its kindred subspecies.

When Ice Age glaciers melted and seas rose, ancestors of the Key deer were stranded on the island chain and developed their distinctive size and an unusual tolerance for salt. They eat mangrove leaves and more than 160 other plants in their luxuriant habitat. Powerful swimmers, Key deer cross channels between islets to find ponds in dry seasons.

OVERLEAF: *"As swift as the roes upon the mountains," a buck bursts from cover, flicking the tail that named his species. He will dash to leave pursuers behind, then hide again.*

The whitetail's antlers sweep forward, single points branching up from the main beams. The rack of the mule deer grows more vertically and each beam has twinned branches.

because their coats are thicker than those of southern races. Tubular hairs give buoyancy for swimming. The coats' insulation enables the deer to lie on snow and not melt it. Most whitetails have reddish coats in summer.

Whitetails, also known as Virginia deer for the area where they were scientifically described, weigh some 200 pounds (91 kg) and measure about 6 feet (183 cm) from nose to tail. They forage on a wide variety of vegetation, including twigs, fungi, and shoots.

In a ritual of recognition, a fawn nuzzles its mother, then begins to suckle. Like other deer, whitetails mate in autumn. Does give birth, usually to two young, in spring. Triplets (opposite) occasionally occur; they survive when deer are well fed and in a relatively safe place—here, a game farm.

A spotted coat camouflages the fawn in shade-dappled thickets. For a few weeks after birth it hides, withholding feces and urine. Then the mother ingests what the fawn voids, denying predators a telltale scent. The young grow quickly, fed on milk with three times the protein and fat of cows' milk.

Moose (*Alces alces*)

Least social of antlered species, this largest of the deer tribe leads a solitary life in woodlands of the north. The moose's name comes from the Algonquin for "he cuts or trims smooth." That moose do, browsing on twigs, leaves, bark, and shrubs. In summer they wade far out into ponds for water lilies. A massive head dips underwater so rubbery lips can pluck plants from the bottom. A big, sensitive muzzle sorts out, by feel, different foods. Bulky, high-shouldered bodies and long legs give moose an ungainly

look. Yet they can speed through water—or snow—while pursuers flounder. Bull Alaskan moose may tower 7 feet (213 cm) and weigh up to 1,800 pounds (816 kg). An exceptional rack of antlers may weigh 70 pounds (32 kg).

Males in rut "nasal test" females (as a bull apparently does above). Although males engage in head-to-head shoving matches for a mate, they are not always the pursuers. Cows lure them with grunting, mooing love calls, which hunters imitate with birch-bark horns. After eight months' gestation the mother gives birth to one young—

sometimes twins. They run with her the first year.

A young moose walks safely before a mother formidable enough to keep a wolf at bay. Moose mothers are so aggressive and attentive that newborn do not have—nor do they need—protective spots.

OVERLEAF: *Wary eye cocked to a photographer at work, a moose cow comes up streaming water. Moose may dive as deep as 18 feet (5.5 m) to feed on pondweed and water lily. Nostrils close during dives.*

Caribou *(Rangifer tarandus)*

In waves and in parallel lines like iron filings shaped by a magnet, the caribou move across the tundra. Their feet click as they move—like castanets—a characteristic shared by all caribou races. A group spurts ahead to a patch of green and pauses to graze. Others trot on. A stream may cause a bunching until some cow or bull takes the lead and swims across.

Thus do caribou of the tundra make their twice-yearly migrations: in one direction to calving grounds in the spring, then back again to winter ranges in the fall. Pastures may be 900 miles

(1,450 km) apart. Most herds number in the thousands.

The species *R. tarandus* is split into a number of subspecies. Two of these, traditionally known as "barren-ground caribou," are *R. t. groenlandicus* in Canada and *R. t. granti* in Alaska.

Caribou feed on sedges, grasses, forbs, and willow and birch leaves. But lichens are the mainstay of their diet. They eat about 12 pounds (5.5 kg) a day, nibbling at the slow-growing plants with small,

- R. t. caribou and others
- R. t. groenlandicus
- R. t. granti

weak teeth. To find enough forage, they must be constantly on the move. Most of the year they roam in small bands of like sex or in loose herds of several hundred; these come together for migration.

Male caribou stand 4.2 feet (128 cm) at the shoulder and average 240 pounds (109 kg). Precocious calves are born singly but dropped within a few days of each other by all cows in the herd.

Spring calves stay close to cows as caribou begin fall migration. This herd totals more than 100,000.

Woodland Caribou (*Rangifer tarandus caribou*)

This caribou lives up to its name, dwelling in boreal forests of aspen, spruce, and jack pine. Within this realm it migrates from boggy fens to drier ridges; the caribou of the tundra usually winter below the Arctic tree line, wandering from stunted growth to open tundra. Caribou of the mountains migrate up and down, from high alpine tundra to low woody elevations. All these types—and the partly domesticated caribou, or reindeer, of northern Europe and Asia—belong to a single widespread genus and species, *Rangifer tarandus*.

Woodland and tundra caribou can be identified by their antlers. The woodland has flatter antlers with many short points. It is also larger and darker. Bulls of the Osborn caribou, largest of the genus, weigh up to 700 pounds (318 kg).

The woodland caribou's habitat makes it less gregarious than the tundra dwellers. Groups are small. In the rutting season, bucks may mate with up to a dozen or so does. The bucks shed their antlers shortly after the autumn rut; the antlers begin growing again in March. Females grow antlers in late summer and keep them until spring calving. This growth schedule gives pregnant does weapons and a chance for dominance as they dispute with young bucks over forage.

Like others of its kind, woodland caribou combat northern cold with an outer coat of long guard hairs and a fine, short underfur. Hollow guard hairs give added insulation and buoyancy; caribou are excellent swimmers. Hair covers nearly all the body of most caribou. The caribou of the tundra shed so much in summer that the discarded hair of a big herd piles up in windrows along the edges of lakes and streams.

Although numerous in some northern areas, this subspecies has become rare in the more southerly afforested parts of its range. Only a few hundred survive in Canada south of the St. Lawrence River, and the single remnant herd in the lower 48 states (in Idaho and Washington) contains only a few dozen.

Peary's Caribou *(Rangifer tarandus pearyi)*

On the island-dotted edges of the far north live the smallest of the caribou tribe—Peary's caribou, named for the explorer who encountered them within 500 miles (800 km) of the Pole. They roam a land where scant food resources limit an animal's size.

This subspecies is now critically endangered because of human environmental disturbance, overhunting, and hybridization with other subspecies. Within recent decades, numbers on the Canadian Arctic islands have declined by about 90 percent to around 3,300 individuals.

☐ Peary's Caribou
☐ Woodland Caribou

Peary's caribou is slightly smaller than its cousins of the tundra, and its winter pelage is paler in color. Like the other subspecies of tundra and woodland, it scoops through snow with a forefoot to reach the vegetation beneath.

Caribou hoofs form a broad digging tool, an efficient paddle, and body support for soft terrain. Dewclaws are large and low so they carry some of the animal's weight.

In summer, horny hoof edges are worn down, and fleshy inner pads are exposed for walking on marshy ground. In winter, the pads shrink and hair between the hoofs grows longer. The rims of the hoofs expand to provide traction on ice.

The clicking of caribou feet can be heard 30 yards (27 m) away. Grazing companions find these reassuring sounds, because they are impossible for predator wolves to imitate. The clicking may be caused by snapping tendons or moving bones.

Family Antilocapridae

This family's name—*antilo* for antelope, *capri* for goat—misnames the family's sole species, the pronghorn. It is neither a true antelope nor a true goat. It is the last remnant of a group of bizarre spiral-horned and fork-horned mammals that arose in North America in the Eocene epoch. Before settlers arrived, pronghorns dotted the plains in the millions. Far fewer now roam from southern Canada to northern Mexico, with greatest numbers in Montana and Wyoming.

Along with deer, bison, and other ruminants, the pronghorn has a four-part stomach, which acts as a fermentation vat. After eating, the animal regurgitates the softened cud for chewing, swallowing, and final digestion. Both sexes have horns (below), their sheaths shed annually. Specialized skin over a bony core produces first hair and then horn. Thus each year the "horn skin" renews the core's outer sheath.

Horn Sheath
"Horn Skin"
Bony Core

Pronghorn (*Antilocapra americana*)

A biologist who had been an army officer discovered that pronghorn bucks were good strategists: They staked out territories with short, easily defended borders. He also found that when the rut began, the complex, rigid system broke down and dominant bucks vanished—

along with does in heat. The biologist found that dominant males had hiding places where they held females for mating. Young bucks could not find them. And, no trysting place, no does.

When courting, a male offers a female a sniff of glands under patches of black hair below the ear (as above). The glands give off a

powerful secretion that helps to identify him and mark territory.

The biggest pronghorns, adult males, stand about 3 feet (91 cm) at the shoulders and weigh 90 to 150 pounds (41 to 68 kg). Bucks' horns average 12 inches (30 cm); does' are shorter than their ears

and lack "prongs." Some females have no horns.

Does breed at about 16 months of age and usually bear twins. The young grow rapidly. When only four or five days old, they can outrun a man.

A high reproductive rate has helped the pronghorn recover from

a low of about 15,000 in 1910 to nearly one million today. A dweller of the dry plains and semidesert, it combats cold and heat with its body hairs. They are hollow and can lie flat to insulate or be lifted to let air circulate. The white hairs of its rump patch can be raised as a

warning signal visible two and a half miles away.

OVERLEAF: *Pronghorns, fastest North American mammals, race through a speed-blurred world. They can run for short spurts at over 50 miles (80 km) an hour and cruise at 25 or 30 (40 or 48).*

Family Bovidae

Bison stringing through Yellowstone snows (right) . . . mountain goats clinging to a lofty crag . . . Jersey cows munching in a dairyman's barn—all belong to the family Bovidae. This great array of hoofed mammals provides us with some of our most spectacular wildlife and some of our most valuable farm animals, domesticated about 8,000 or 9,000 years ago.

Almost all bovids have horns, usually borne by both sexes. Dewclaws normally are small or missing. Members of most species live gregariously. Bison and muskoxen exhibit contagious behavior in grazing and movement. All bovids are ruminants, able to regurgitate food and chew it a second time. Those that feed on grass pull it rather than bite it off, to get at the tender lower stems. Although most family members are grassland dwellers, some species are found in desert, mountains, and tundra.

☐ Bison's Historic Range
■ Wood Buffalo Herds
☐ Park and Refuge Herds

In three hours this brand-new bison will run with a species that almost died. Today bison roam in protected herds. For preservation purposes, Canada has isolated its wood buffalo. They had been breeding with plains bison and disappearing as a race.

Bison *(Bison bison)*

A large bison bull may stand six feet (1.8 m) at the shoulder and weigh a ton (0.9 t). His massive weight is concentrated in his forequarters, his heavily muscled neck supports a low-hung head, and his matted forelock hair forms a thick shock absorber known to have stopped bullets.

A bull can push a one-ton opponent backward 10 to 15 feet (3 to 4.5 m). And when two bulls charge, the spectacle is awesome. Usually they first roll in shallow wallows—horn-dug through the sod, the subsoil churned to flourlike dust. They posture and paw the ground. And then the dust explodes as they lunge at each other from a few feet apart— and collide.

Deep, bellowing roars accompany such fights. John James Audubon, camped on the plains in 1843, noted that the din of the numerous conflicts sounded like "the long continued roll of a hundred drums."

Hooking horns rip out hunks of hair. Wounds—and death—occur. But usually the clashes end when one bull turns his head or body aside in submission, recognition of the rival's dominance.

Fights occur mostly during rut in late summer and early fall. Until then bulls have wandered alone or in small groups on the periphery of larger bands of females and subadults. The bands roam together loosely as a herd.

In rutting season, the bulls join the bands, busily competing with rivals and tending one cow after another until each is mated. A bull

may lose 300 pounds (136 kg) during the rut. Calves are born in spring after a gestation of about nine months.

Adult animals lose some of their distinctive foreleg "pantaloons" and other display hair after the rut. Sexes then are quite similar in appearance, though females are smaller—averaging 5 feet (152 cm) and 930 pounds (422 kg)—with slenderer horns whose tips point forward. (Bulls' curve upward.)

Plains bison, misnamed buffalo by European settlers, use their ranges erratically. No firm evidence supports former belief that they migrated extreme distances north and south with the seasons. Herds may seek winter shelter of tree belts in storms, or move to better forage in snow—sweeping muzzles like brooms to dig down as deep as four feet (1.2 m) for food.

Once bison ranged almost all the continent. To Plains Indians they were a four-legged commissary, yielding "meat, drink, shoes, houses, fire, vessels, and their Masters whole substance," an explorer wrote. Hunters on the plains after the Civil War virtually exterminated them. Today plains bison number about 200,000.

The WOOD BUFFALO (*Bison bison athabascae*) exists in two small herds in Canada. It is larger, darker, and more wary than the plains bison.

OVERLEAF: *Few where once there were millions, bison thunder across plains their ancestors ruled. Bison can run 32 miles (51 km) an hour in short bursts.*

Mountain Goat *(Oreamnos americanus)*

Mountain lions may corner it in a cave when it seeks shelter at night or in a storm. Wolves may pounce on it when it descends to wooded valleys and alpine meadows. Eagles may swoop down and seize its kids. But when the mountain goat takes to its lofty realm, few foes dare to follow. (Safely gamboling on the high pass above are, from the left,

an adult female, two adult males, and a leaping youngster.)

The scientific name suggests that the Greek Oreiad, a nymph of the mountains, has been reborn in America. But the common name offers a zoological error, for *O. americanus* is not a true goat. It

belongs to the goat-antelope group that has adapted to life on the crags. It is kin to the chamois of the Alps and came here from Eurasia in an early ice age. Up to 100,000 now can be found in remote high country in northwestern states, Alaska, and western Canada, where most live.

The mountain goat has a body built for climbing: flexible hoofs; compact, muscular torso; rather short legs—poor for running but ideal for balance.

Males and females, including yearlings, have stiletto horns that can deeply puncture even the tough skin of the male's rump shields, which thicken during rut. Natural selection has led to aggressive, horn-wielding females that zealously guard their kids and dominate males except during rut.

Courting males crawl on their bellies and squeak tenderly like baby goats to win a nanny. After mating, the billy prudently leaves— or gets chased away.

On a diet of grasses, sedges, and shrubs, males grow to an average 3.5 feet (107 cm) in shoulder height and weigh from 190 to 280 pounds (86 to 127 kg). Females weigh about a third less.

355

Muskox (*Ovibos moschatus*)

Despite *ovibos*—"sheep ox"—and *moschatus*—"musky"—of its scientific name, this animal is neither sheep nor ox nor musky. Its misnomer may stem from the odor of urine on bulls in rut; muskoxen have no musk glands.

Confronted by an intruder, they move with an instinct born of the ages. The herd forms itself into a defensive ring (as above), massive heads facing out, calves protected in the center, behind the furry rampart. If the intruder steps closer, an adult will rush out to meet the threat with a goring or trampling.

Such tactics, developed through millennia of battle with wolves, worked against those predators. But that behavior meant suicide in contests with human hunters, especially the ones bearing guns. An entire herd of muskoxen might be shot where it stood.

Once these Ice Age relics roamed with mastodon, hairy mammoth, and woolly rhinoceros, but they had been exterminated in Eurasia in prehistoric times. In North America in 1689 Henry Kelsey became the first of many explorers to see the "ill shapen beast. . . . their Hair is near a foot [30 cm] long."

In the 1850s Alaska saw its last muskox killed. Thousands in Canada were slaughtered to meet demands of the lap-robe trade.

Muskoxen were so rare by 1917 that Canada ordered total protection. Their numbers have since recovered to over 50,000; herds have been reestablished in some former ranges.

During the rut, bulls that have been solitary or in bachelor bands rejoin groups of females and younger animals. Herds usually average around 15 members and may number as many as 100. Bulls are polygamous and compete for mates. Two rivals may charge each other repeatedly in head-on clashes until one concedes.

Though clumsy looking, muskoxen are nimble. Bulls measure about 4.5 feet (137 cm) at the shoulder and weigh an average of 750 to 800 pounds (340 to 363 kg). Females are smaller and their horns are less broad.

Muskoxen roam river valleys, lakeshores, and damp meadows in the summer, feeding on willows, sedges, and grasses. They do not particularly like lichens, the caribou's chief food, and so the two species can share the same range. In winter, muskoxen move to hilltops, slopes, and plateaus, where the wind helps clear the vegetation of snow. They survive on a sixth of the fodder needed by domestic cattle.

Bitter cold following wet snow may turn the crust to concrete hardness; unable to paw through to forage, muskoxen starve.

Dangling guard hairs—as long as 24 inches (61 cm) on a bull's neck—shed moisture. Underneath, a dense, downy layer of wool finer than cashmere turns away cold. Shed in April or May, it trails after the animal in streamers, or clings in patches to tundra shrubs and plants.

To Eskimos the animal was *Oomingmak*—bearded one. They wove mosquito nets from the guard hairs. *Qiviut,* they call the gossamer-light, incredibly warm underwool. A pound (500 g) of it can be spun into a strand 10 miles (16 km) long; four ounces (120 g) is enough to knit a dress. A muskox yields about six pounds (2.7 kg) of the wool, and for it ranchers hope to domesticate the adaptable, mild-tempered animal.

Dall's Sheep (*Ovis dalli*)

Wary when hunted, Dall's sheep will flee at a hint of danger. But where they are protected they will accept and even approach humans. (The males above cluster to stare inquisitively at the photographer.)

The wild sheep's ability to get used to human beings may explain why sheep became domesticated very early in our prehistoric past.

Dall's rams banding together demonstrate a trait of mountain sheep behavior. After the rut, males leave herds of ewes and juveniles to roam in small bands. They move from one grazing patch to another in a pattern set during interglacial periods, when climate changes shrank the grasslands their ancestors roamed.

Good grazing patches may be 40 miles (64 km) apart, and elders know the routes through the valleys that stitch the patches together. Young males cannot go off on their own because they have not learned the way. So they must tag along.

But how does a youngster stay out of harm's way among big, aggressive rams? He imitates females. This behavior redirects aggression by the dominant rams, allowing the smaller, weaker animal to stay in male company.

Stone's Sheep (*Ovis dalli stonei*)

Stone's sheep, a thin-horn, has been called "a Dall's sheep in evening dress." The white of rump, belly, head, and leg trim contrasts with the silver gray through brown to near black of its body.

As with other mountain sheep, a Stone's ewe bred in autumn's rut bears her lamb in spring. She seeks a cliff or the isolation of broken country to give birth and there cement maternal bonds. In an experiment, ewes failed to pick their own from lambs tied in bags and presented headfirst. But when the bags were reversed, the ewes had no problem. Anal glands made the difference. In a herd, a mother will nurse only her own. But a female with no lamb of her own may become what author Geist calls an "aunty" that young flock around.

Young males join roaming ram bands after their second birthday. Young females stay with the herd.

Stone's head-to-tail measure of about 5.3 feet (162 cm) makes it longer-bodied than Dall's sheep. It also is somewhat heavier.

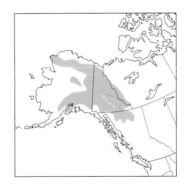

■ Dall's Sheep
■ Stone's Sheep

Dall's sheep belong to the "thin-horn" branch of *Ovis*. Their horns spread and spiral more than the bighorn's; tips point away from the face. Males average about 3 feet (91 cm) in shoulder height and weigh from 165 to 200 pounds (75 to 91 kg). Grasses and sedges form their principal food.

Bighorn Sheep (*Ovis canadensis*)

Among *Ovis* species, horn size forms a symbol of rank. In the dominance hierarchy, the bigger the horns, the more dominant the owner. A ram's horns may weigh 30 pounds (14 kg), as heavy as all the bones of his body. The horns of an old ram (above) can be a biography. The horns are blunted and frayed at the tips, evidence of repeated bouts with his rivals. Ringing the horns are deep grooves—marks formed each autumn when annual growth stops. These show him to be about 10 years old; 14 is long-lived. Segments between creases narrow toward the base as growth slows with age and as each increment pushes older growth outward.

Butting jousts may occur all year but intensify during rut when rams join herds and vie for mates. Two rivals will rear on hind legs, then drop to all fours and slam together head on. The shock waves ripple through their bodies. Recovering, each freezes for up to a minute in a show-off of horns. Thus the blow received is instantly equated with the size of the horns that inflicted it.

For battle protection, mountain sheep have double-layered skulls shored with struts of bone. A broad, massive tendon linking skull

Suckling bighorn depends upon its mother for sustenance. But it assumes independence early, going off to gambol with other lambs. Play strengthens its limbs and body for life on the ledge.

and spine helps the head pivot and recoil at the blow.

A ram may weigh 250 to 280 pounds (113 to 127 kg) and measure 5.5 feet (168 cm) from head to tail. It can smash into an opponent at 20 miles (32 km) an hour.

O. canadensis is found in the Rocky Mountains from southern Canada to Colorado, and as desert subspecies south into Mexico. From an original population in North America of over one million, only some 60,000, or less than 5 percent, remain today. The bighorn's social traditions dim much hope of recovery.

Unlike moose and white-tailed deer, whose young disperse and colonize new areas, mountain sheep young follow their elders. The bighorn is also imperiled by poachers for meat and horns, by diseases, by competition from livestock, and by continual human encroachment on habitats.

Adults seem to adjust to such pressures by holding down populations. The more restricted the habitat or the more scant the forage, the less viable the young and the smaller the herd.

The predominant bighorn of the desert, Ovis canadensis nelsoni, is smaller than its northern cousin and has flatter, wider-spreading horns.

The desert bighorn can be found in slightly varying forms from Nevada and California (left)—with a few in west Texas (above)—into Mexico. It inhabits hot, dry, rocky, cliff-and-mountain regions where its leanness gives it a large surface-to-weight ratio beneficial in getting rid of body heat.

Like other mountain sheep, desert bighorns aren't as well built for clambering as mountain goats. But bighorns can zigzag up and down cliff faces with apparent ease. A two-inch (5-cm) ledge is enough for a foothold, and a spot too small for a stand may be a place for a pause in a bounce from niche to niche over spans as wide as 20 feet (6 m).

These bighorns get moisture from desert plants. Though they need water, it may be several days between visits to water holes.

Rams stand 35 inches (89 cm) at the shoulder, weigh about 200 pounds (91 kg). Their rut season—July to October—is the longest of any mountain sheep. But single young and social traditions keep populations small. Competitors in the desert—wild burros and people—threaten the sheep's future.

OVERLEAF: With a crack so loud it can be heard a mile (1.6 km) away, two rams clash in Alberta's wilds. Rivals may butt repeatedly in a stylized ritual. Author Geist watched a determined pair collide some five times an hour for $25\frac{1}{2}$ hours before one finally conceded.

Alien Mammals

Introduced Species

Rhesus monkey business along Florida's Silver Springs River goes on in social groups, as it would in native India. The feral animals are descended from unemployed 1930s movie monkeys.

IN JANUARY 1494 Christopher Columbus wrote home from Hispaniola. Half his men were sick, he told his sovereigns. "Under God," he said, "the preservation of their health depends on these people being provided with the food they are used to in Spain." Ferdinand and Isabella sent supplies and livestock. Such shipments from Spain began the introduction of exotic—nonnative—animals to the New World. Some survive only as domestic stock. Others have established wild populations which often endanger native plants and animals. Although today we try to control or forbid by law the importation of foreign species, our past record has been thoughtless and often catastrophic.

Columbus supposedly left swine on Hispaniola when he went home in 1496, and their descendants, transported to the mainland, were probably the first pigs in North America. In 1521 hogs were penned aboard the ships that Juan Ponce de Leon sailed to Florida. Hernando de Soto, in 1539, brought 13 hogs to the North American mainland. According to expedition journals, they "increased to three hundred swine." Escapees from Spanish herds were perhaps ancestors of the wild hogs that are now found in the southeastern United States.

The horse, unlike the pig, was here long before Columbus, as *Hyracotherium*—often known as *Eohippus,* "dawn horse" of the Eocene—some 55 million years ago and, about three million years ago, as its descendant, *Equus,* the genus to which the modern horse belongs. *Equus* prospered until about 8,000 years ago when, along with other large mammals, it vanished from the continent. Ecologist Paul S. Martin believes that this massive extinction was not the result of climatic changes but the work of Stone Age hunters who slaughtered many species out of

existence. "The complete removal of North American horses," he says, ". . . represents the loss of a lineage of grass-eaters, without the loss of the grass! It left the horse niche empty for at least eight thousand years."

The Spanish brought this lineage of grass eaters back. From breeding farms in the West Indies conquistadores picked up pack animals and fine mounts. Francisco Coronado had more than 500 with him when he encountered Plains Indians in 1541.

At first, the Indians probably feared the horse. But, a Cree warrior said, "as he was a slave to man, like the dog, which carried our things, he was named the Big Dog." Although no one knows when the first mounted Indian proudly rode off on his Big Dog, horse ownership grew among the tribes through trading and raiding, and by 1800 the Plains Indian was transformed from a foot stalker to a swift and skillful mounted hunter.

Another *Equus* species, the burro, followed the conquistadores and eventually scattered to the wild after having served as beast of burden all over the American West. Today both horses and burros roam in feral herds. Are they "living symbols of the historic and pioneer spirit of the West"? So says the 1971 Act of Congress that protects those on public lands from "capture, branding, harassment, or death." Or are these animals ecological misfits, intruders in niches belonging properly to other wildlife?

By the 1960s the native bighorn sheep had almost disappeared from several areas of California because, state biologists believe, the booming burro population competes with the sheep for forage and water. A game official accuses burros of "competing with every living animal on the desert, right down to salamanders at the water holes"—where burros trample the ground enough to destroy

The ibex (right), a mountain goat hunted for its horns in its native Siberia, now leaps canyons and cliffs of the Southwest.

An Asian elk often called a deer, the sika thrives in feral herds in Maryland, Virginia, and Texas (opposite). Sikas were first released in 1916 on Maryland's James Island. Some of these able swimmers crossed Chesapeake Bay to the mainland. Others, introduced to Assateague Island, prosper in a herd of about 1,000.

habitat for many kinds of small animals. At Grand Canyon National Park, rangers shot 2,800 burros between 1924 and 1969 to protect native plants and animals. The burros that remained in the canyon increased by about 20 percent a year. To ease ecological damage, nearly all have been removed.

Wild horses also overpopulate and overgraze, in the view of people who decline to see the western states as a vacant niche for the feral horse. It is often charged with usurping

cattle range—as "America's number-one interloper." It has long been treated as a pest. From the early 1900s, mustangers, now outlawed, rounded up horses for the hide buyer, and soon after for chicken-feed and pet-food factories.

Wild horses live in remote, rugged country. In 1925, before they could be counted accurately, they were estimated at about one million. By 1971 the Bureau of Land Management guessed that only about 17,000

ranged public land in nine western states. Then animal lovers saw the fruition of a nationwide campaign: the Wild Free-Roaming Horse and Burro Act, which promised to halt the slaughter. But this, in turn, has raised other controversies.

In 1994 more than 36,000 wild horses (and some 6,000 burros) grazed western rangelands. But overcrowding has been eased, as has competition with native wildlife and domestic stock, by the Adopt-A-Horse program. Between 1971 and 1994, about 135,000 horses and burros had found new homes with citizens in all 50 states.

Into the arid Southwest, one day in 1856, strode 34 camels, brought from the Near East to serve the U. S. Army as pack animals. They turned out to be touchy, independent creatures, biting, kicking, and spitting at the mule skinners who mistreated them. They spooked the cavalry horses and frightened citizens. Even so, more were imported, some by civilians. Eventually the unfortunate camels were sold or turned out to fend for themselves. They wandered the desert until at least 1905, occasionally producing young, and turning up now and then, like phantoms, in a mining camp or a ranch yard.

Not all our exotic species were invited. Most unwelcome are the Old World rats and mice. Native to Asia, all live near humans. As fellow travelers, the house mouse and the black rat had spread to Europe by the Middle Ages. Then in the 16th century they scrambled aboard ships to the New World. They put ashore on America's east coast and soon also on the west.

By 1825 the black rat had become so well established on the east coast that it was taken to be a native mammal. The Norway rat appeared in the American colonies about 1775—the English claimed the colonists had brought it in with smuggled goods, and the colonists blamed the English, while *Rattus norvegicus* went about its own colonizing, covering the continent.

Worldwide, rats and mice eat and contaminate about a fifth of all food crops planted. In the United States alone, in a year's time, they damage or destroy property worth more than $1 billion. They gnaw wiring and cause innumerable fires. Through fecal droppings, they transmit several diseases to humans. Their parasites carry such diseases as murine typhus and plague—the "black death," which killed an estimated one of every three Europeans between 1347 and 1350. Plague is transmitted by fleas among rats and other small mammals. If its animal host dies, the

flea may move to and infect a nearby human.

Rat-borne plague struck San Francisco in 1900 and Los Angeles in 1924. The disease smolders here in such wild animals as prairie dogs, rabbits, and squirrels. An average of ten people contract plague every year, mostly in the western states.

But the house mouse and the Norway rat atone for some of the sins of their kind: The albino strains of both species are the mice and rats of medical laboratories. And biologists tell

Not quite at home on the range, these imported ungulates roam commercial game ranches in Texas. With more than 30 other species native to Africa and Eurasia, axis deer (opposite), oryx (above), and blackbuck antelope (left) lure hunters from all over the world. Sometimes the "Texotics" escape their fences. No one yet knows whether the free-ranging grazers and browsers will adapt to a new habitat in harmony with native species. The Texas game laws do not protect most exotics. In any season a handsome rack can become a trophy on a "no kill, no pay" safari.

371

us that "dirty rat" is slander, that laboratory rats quickly rid themselves of parasites, and that rats in the wild groom daily and choose clean food. It is the environment, which we untidy and wasteful humans provide, that makes the rat our enemy.

By far the greatest number of nonnative mammalian species have been brought to North America for the guns of hunters, and often with little forethought. A particularly bad choice was the Old World rabbit. In spite of its infamous past—this rabbit ravaged Australia in the late 1800s and thousands of miles of fence failed to stop it—it was released in the Midwest in the 1950s for hunting and for running in beagle field trials.

The animals came from the San Juan Islands north of Puget Sound, where they had lived since about 1900 and where their multitudes had eaten plant cover to the ground and burrowed so extensively that the bluffs of one island crumbled into the sea. Luckily the introductions in the 1950s and others later seem to have failed.

In the Great Smoky Mountains they tell about wild "Rooshians," the boars imported there in 1912 to a private game preserve for sporting gentlemen. From Russia—or Germany, depending on who's telling the story—came 14 wild boars, to be penned in a great enclosure so that they could "increase in number." On the first hunt, a few years later, the powerful hogs knocked down the fence and escaped into the woods. Interbreeding with feral domestic swine, they have increased in number ever since.

Wild boars did much damage in the Great Smoky Mountains National Park, eroding soil by their rooting, destroying habitats of small animals, and stripping plant cover. More than 6,000 boars have been removed since 1977. Today, over 500 still range the park, and half the adults must be taken out each year to keep their numbers stable. Trapped boars are usually released on state lands for hunting.

But in the Southwest, a wildlife manager justifies the importing of game animals by citing "our American tradition that every citizen should have the right . . . to hunt." Another suggests "limiting the kill to the most admirable and challenging creatures." Some state game commissions have endorsed the careful introduction of exotics to land depleted of native game. Most importations, however, have been made by private landowners, who cite their success in providing sanctuary for some species that are threatened with extinction in their native lands. The exotic-species count today—highest in Texas—is about three dozen.

Ranchers have gone into the game business in a big way. Hunters pay fees to bag exotic trophies—up to $1,500 for an axis deer or a Barbary sheep, $4,500 for an oryx. On many ranches, the income from hunting exceeds that from livestock. Usually "Texotics" are fenced. But there is always the chance that a species will establish a wild breeding population and threaten native wildlife.

The European hare was also imported for sport hunting. By the 1920s it had made its mark as a clever quarry—and a pest to farmers. Another small mammal, the nutria, brought from South America to Louisiana as an experiment in breeding a new furbearer, became a major money crop for the trapper.

The haphazard history of the nutria in Louisiana began in 1938. Several pairs were imported from Argentina and placed in "escape-proof" pens at Avery Island. Here they multiplied, prospered—and escaped, many during a 1940 hurricane. More were imported. By the late 1950s Louisiana counted 20 million nutrias nibbling away at its wetland. The pelt supply zoomed. The market slumped. Then in 1957 Hurricane Audrey further destroyed the vegetation cover that the nutrias themselves had damaged, and in one bad winter millions of the animals froze to death. Their population stabilized, and the native muskrat population declined. Prices for nutria skins rose, and Louisiana found itself with a multimillion-dollar nutria industry. But what happened was all a game of chance. To protect our native plants and animals, we must stop taking chances.   STEPHEN R. SEATER

Horse (*Equus caballus*)

Strong legs and the hoof (that single digit upon which the horse runs) give *E. caballus*—agile, fleet, ever on its toes—the most advanced limb structure in the order Perissodactyla. Wild horses roam pockets of British Columbia and the western United States. To most eyes, they are a motley crew, the Spanish blood long ago diluted by feral horses of many breeds. Color varies. An occasional palomino, black, white, or pinto appears among the brown and bay multitudes. Today's wild horse, while often foraging a meager range, grows to about 5 feet (1.5 m) at the withers.

Though small, wild horses prove their stamina. They can run for miles, spurting to 35 miles (56 km) an hour. Swift flight is the horse's instinctive first defense, though the natural predators it once faced, such as cougars, are gone from most of its range.

Wild horses often travel in herds of 3 to 20 animals—but a herd can be larger. A stallion usually commands, with an older mare as lieutenant. So dominant is a prime stallion that his herd is known as a harem and he furiously fights off any rival. To drive his mares, the harem master lowers his head, stretches his neck, and lays back his ears. Dawdlers get a sharp nip on the flank. When the horses graze or drink at a water hole, the stallion stands guard.

Mares go off on their own to foal. Gestation takes 11 months, and the young usually are born in the spring. Colts stay with the herd until, at about three years old, they show signs of growing up. Until the age of six or seven, they roam (like the young stallion above) with other bachelors. By raiding or by deposing an old stallion, they get their own harems.

OVERLEAF: *Wild spirits flee, spurred by a dappled stallion as rearguard. When a herd takes flight, a shrewd old mare chooses the escape route.*

Black Rat (*Rattus rattus*)

A climber and high-wire artist, using its long, scaly tail as a balancing pole, the black rat seeks lofty places. It colonizes attics and rafters of barns and buildings—thus its other name, roof rat. It seldom descends to burrows, basements, or sewers. Outdoors—in affluent Florida and California neighborhoods, for instance—this "bare-tailed squirrel" nests in trees.

R. rattus once disembarked daily at ports in the U. S. and Mexico. Modern ratproof ship construction and inspection methods have virtually blocked that route. It lives along the south Atlantic and Gulf coasts and on the west coast north to British Columbia.

Seldom does the black rat weigh as much as a pound (0.5 kg) or exceed 18 inches (46 cm) overall.

Its slender build, longer tail, and naked, prominent ears distinguish it from the larger Norway rat.

Companion and foe of humans from earliest times, *R. rattus* lives by devouring immeasurable amounts of our food. The settlers at Jamestown, wrote Captain John Smith in 1609, found their supply of cracked corn "so consumed with the many thousand rats, increased first from the ships, that we knewe not how to keepe that little wee had. This did drive us all to our wits ende."

At wit's end is where we often are today, trying to cope with the proliferation of these rodents. Some researchers say that traps and poisons only make room for more. The rats breed year-round, with five or six litters a year and up to seven young per litter. Offspring begin to breed by the fourth month. Mortality among young rats is high, but a pair theoretically could produce 1,500 descendants in a life span of one year.

Norway Rat (*Rattus norvegicus*)

"The finest . . . product that Nature has managed to create" is probably the Norway rat, wrote naturalist Ivan T. Sanderson. Adaptability—both behavioral and genetic—is perhaps the most distinguishing trait of *R. norvegicus*.

In every state, in Mexico, and in southern and western Canada, wherever people have concentrated, and in rural areas nearby, the Norway rat lives. Although in temperate climates it may summer in grainfield or haystack and winter in barn or silo, this rat prefers human density, with its food waste and diversified shelter. A burrower, it lives on ground floors, in basements, tunnels, garbage dumps, and—since it is also a great swimmer—in sewers.

Of heavier build and with a shorter tail than *R. rattus,* the Norway rat has reddish gray to brownish gray fur that gives it another name—the brown rat. In spite of tales exaggerating its size, even the largest weigh less than 2 pounds (0.9 kg).

Norway rats eat almost anything, though they prefer grains, meat, eggs, fruit, or a chance to sort out the garbage. Rats sample all new food cautiously. If bait tastes strange or makes them ill, they will become bait-shy and reject it forever.

They live in hierarchical groups. The dominant males, their females, and offspring nest closest to the food supply. Again and again, the strongest survive to breed most successfully. Their high breeding rate also has allowed the rapid spread of genes that provide resistance to anticoagulant poisons, thus producing a "super rat" that, so far, defies our attempts to eradicate it.

House Mouse (*Mus musculus*)

Anything we eat, *M. musculus* eats. Biologists call such a species a "human commensal," an animal that shares our table—and usually our buildings. But the house mouse also nests in meadows, grainfields, and sand dunes. Any place in North America where people live or visit, it makes its home. Its length, including tail, is from 5.1 to 7.8 inches (130 to 198 mm).

The tiny size of the buff-bellied, gray-brown rodent is, for it, both a curse and a blessing. It is prey to larger meat-eating animals. But the

mouse can hide in the smallest crevice or hole, under any leaf or rock, in cupboards, drawers, and coat pockets.

We often feel affection for this tiny animal—"little mouse" is a term of endearment. In folk belief, a mouse was the form taken by the soul as it escaped from the mouth at death. And mice "sing" in the night, twittering and churring.

Yet we know that this mouse is a destroyer and contaminator of food and water, a vector for some of the diseases carried by the infamous rat, and a pest worthy of pursuit by our folklore farmer's wife. In cities

where poisons and improved sanitation reduce rat populations, the mouse menace increases in proportion. Their food, water, and range needs are limited, and they show greater resistance to anticoagulants.

Like most rodents, *M. musculus* is prolific. Females tend to come into heat in the presence of males and produce from 5 to 14 litters a year. (The one above is newborn.)

Each litter of 3 to 12 may breed in 6 weeks.

House mice colonize. Dominant males establish both a home range, clearly marked by odor (probably urine), and a family of several females and young. Members of foreign groups are attacked and cast out. This ensures some genetic isolation, reinforcing genes adapted to the environment. If a community grows too large, groups of young leave to form new colonies. Thus *M. musculus* exploits genetics and chance in the battle for survival.

Nutria (*Myocastor coypus*)

Its thick fur makes a fashionable jacket worth up to $3,500. But tunneling in dikes and gobbling in sugarcane fields makes an outlaw of this marsh-loving rodent—the nutria, or coypu, as it is also called. The only member of the family Capromyidae successfully introduced to North America, *M. coypus* lives along streams and rivers of southern Canada and scattered areas of the United States from Maryland to the state of Washington. It flourishes in the warm Gulf Coast climate.

A streamlined swimmer, a nutria has a triangular head, small ears that close out the water as it dives, and webbed hind feet. A female bears one to three litters a year, as many as nine young each time. Fully furred at birth, the young can navigate on their own when only five days old. But they usually stay with the mother six to eight weeks. As she swims, they ride on her back and can nurse in transit—the mammary glands are conveniently high on her sides, above the

waterline. Before the young are full-grown they may breed.

An adult is a robust 8 to 20 pounds (3.6 to 9 kg) and about 3 feet (91 cm) long, including a round, scaly tail. Nearly hiding the soft, gray underfur are long, coarse guard hairs of yellowish or reddish brown. Originally, the fur trade prized only the underfur of the belly. As fashion changes, the market also uses long-haired

pelts, as well as sheared or plucked. The best furs now come from Argentina.

Cattails, sedges, and water weeds are favorite foods of the usually gluttonous nutria. A dense population can strip marsh areas of vegetation, creating a wasteland that Louisianans call an "eat-out." A burrower, the nutria dens in banks of rivers or streams, and can honeycomb a rice-field levee. In shallow water, *M. coypus* mounds reeds and other plants (as shown above) for grooming and feeding platforms.

European Hare (*Lepus europaeus*)

Head over heels on a downhill sprint; a straightaway spurt at 30 miles (48 km) an hour; an uphill streak on long hind legs—these are the speed feats that make the European hare a challenging game species. (Sometimes this hare is classified as *L. capensis*.)

These hares—at 25 to 27.5 inches (64 to 70 cm) the largest in their range—occur in Ontario, Canada, and in the northeastern corner of the U. S. They prefer open country where in the daytime they crouch in clumps of grass or brush. At dusk they emerge to feed on grass and herbs in summer, twigs and bark in winter. If discovered by a hawk, fox, bobcat, or a human hunter, they usually bolt, artfully dodging and doubling back on the pursuer. *L. europaeus* has been known to lead dogs across ice thick enough for itself but too thin for them, to bound with ease over a five-foot (1.5-m) wall, and to swim wide rivers.

The kinky, grizzled pelage, brown in summer (above), lightens in winter. European hares are active year-round, and the first young of the breeding season may appear when snow is on the ground. The doe scatters the leverets among nearby nests. When returning to nurse them, she often utters a faint grunt, which they answer. By March she may be pregnant again.

Wild Boar *(Sus scrofa)*

Signs that the wild boar is near: a wallow in a streambed, earth bulldozed far and wide, tree trunks scraped and muddied.

Member of the Old World family Suidae, this boar—long since interbred with feral domestic swine—ranges marsh, forest, and mountain in New Hampshire, California, and southern states.

Few predators vex this giant among pigs, and only the black bear is its match. Built like a small bison, *S. scrofa* grows about 3 feet (91 cm) tall.

If disturbed, boars usually flee, and they are swift, agile runners. But when antagonized, they fight, slashing with deadly tusks. These elongated canine teeth (right) grow continually, sharpening as they wear against each other. With tusks and a disk-shaped snout of strong cartilage, *S. scrofa* grubs for roots, tubers, insects, and earthworms. Sharp incisors tear up grasses, crop nuts and fruit, or chomp on snakes—even rattlers—sometimes killed first by a blow from a sharp hoof. Fungi, leaves, snails, young birds, small animals are all fare for a hungry boar.

Boars usually forage at dawn or dusk, though some are completely nocturnal. Except for breeding, most males travel alone, females and young in small bands.

Cooling off in a mud wallow may consume much of a wild boar's day. Occasionally, *S. scrofa* plucks grass or other vegetation with its mouth, spreads it over a small area, and crawls under it. The cut grass is lifted, interwoven with standing grass—and the burly boar relaxes in the shade of a lacy canopy.

Burro *(Equus asinus)*

"The tattered outlaw of the earth" in G. K. Chesterton's poem, *E. asinus* is many things to many people. The tourist on a Death Valley road sees a cute little moocher. The Grand Canyon hiker once found a fouler of water holes. A park ranger sees a competitor of native wildlife. Ass of the Bible, *el burro* of the Spanish padre, pack animal of the gold rush, donkey of children's zoos—it has been these also.

Descended from animals lost or cast off since the 16th century, an estimated 6,000 burros roam southwestern North America. But they are hard to count in their remote canyons, mountains, and deserts. They are hardy animals of all seasons, have no natural predators, and overpopulate much of their range.

The long-eared burro has a shaggy coat—usually brown, gray, or black and often lighter at nose and belly. It has small, sure feet, an erect mane, and a tufted tail. An average adult weighs about 400 pounds (181 kg) and stands up to 4 feet (1.2 m) at the withers.

A female with young forms the only stable social unit. Jacks may dominate a small territory or wander in changing bachelor groups, breaking off to pursue a jenny in heat (above) or to battle in a melee of brays and hoofs, where the teeth of one male can grip an ear of the other and whirl him in a circle.

The jenny bears a foal—about every year and a half—after a gestation of 12 months. Before it is a week old, the suckling begins to forage with its mother and test its keen nose for water. Whether in alpine meadow or desert wash, burros are not fussy eaters. They crop grasses or shrubs, thriving on plants most browsers reject: A burro can de-spine the prickliest cactus to get the pulp inside.

The Ultimate Mammal

THE MAMMALS CALLED people (the species *Homo sapiens*) are the only existing members of the family Hominidae of the order of Primates. Our species seems to have originated in the Eastern Hemisphere about 200,000 years ago and to have entered the Americas at least 20,000 years ago. For most of this time *H. sapiens* seems to have had little effect on other species. But toward the end of the last ice age, people attained numbers and cultural levels that allowed intensive hunting of the largest land mammals. About 12,000 years ago, such skilled hunters began moving across North America.

From fossil remains we know that during the next several thousand years there was a wave of extinction unsurpassed since the end of the dinosaurs. Mammoths, horses, ground sloths, giant beavers, and about 30 other kinds of large mammals disappeared from North America. Why? Scientists have offered several theories: Perhaps hunters killed so many plant eaters that those species died off, and the carnivores, such as the lion and saber-toothed cat, then perished for lack of prey. Or perhaps hunters merely finished off species already in decline because of deteriorating environments. Or possibly people had only an indirect effect—by disrupting the ecological and behavioral patterns of animals.

Sea mammals were largely unaffected by the catastrophe that engulfed land mammals at the end of the Pleistocene. But by the 15th century, when the second major human invasion of North America began, people had learned how to efficiently hunt sea-dwelling species. And because marine mammals were highly visible and yielded meat, oil, and other products, they were pursued intensively.

Explorers found two species of sirenians at opposite corners of the continent. Meat-hungry Russian sailors wiped out the giant Steller's sea cow by 1768, only 27 years after its discovery in the Bering Sea. The manatee, found in the waters of Florida and adjacent areas, was in trouble soon after. But somehow that species survived until it was protected by state law in 1893. Not so fortunate was its neighbor, the Caribbean monk seal, one of the world's few tropical pinnipeds. Already declining rapidly by 1800, the species lingered in remote areas through the mid-20th century. Recent surveys indicate that the seal is extinct.

Pinnipeds that do not gather in vast breeding colonies usually have avoided excessive human exploitation. Species that do concentrate periodically in small areas have met disaster. Hunters killed so many Guadalupe fur seals and northern elephant seals along the coast of California and Baja California that both species were thought extinct in the late 19th century. Each seal has since been rediscovered, protected by law, and allowed to increase in numbers.

The northern fur seal was easily slaughtered when some 2.5 million gathered each year on a few small islands in the Bering Sea. Several times attempts to control the killing created political crises involving Great Britain, Russia, Japan, and the United States. In 1911, when only 200,000 animals remained, these nations agreed to jointly protect the herds. The seal population increased to about two million, then declined again by more than 50 percent.

One of the most important industries of the New England colonies was whaling. Initial targets included the coastal-ranging Atlantic gray whale, which apparently disappeared in the 1700s, and the right whale. As stocks declined in the 18th century, American whalers turned north after the bowhead whale of Arctic waters and spread over the world's

Victims of costly, controversial predator-control campaigns, dead coyotes dangle from a U. S. Fish and Wildlife Service helicopter. Each year tens of thousands of coyotes are gassed, poisoned, trapped, and shot—many by airborne sharpshooters. These operations are currently conducted by the U. S. Department of Agriculture.

seas in pursuit of the sperm whale. By about 1860 whaling had slackened in the United States, but new developments allowed other nations to intensify their whaling.

Perfection of the harpoon gun in the 1860s led to large-scale exploitation of the swift-moving blue, fin, and sei whales. During the early 1900s tremendous stocks of whales were discovered in Antarctic waters, and soon these were being hard hit by whalers operating from island bases.

The 1920s saw the first factory ships, which fully processed at sea the carcasses of whales killed by crews in smaller accompanying boats. With this innovation came a substantial increase in the take of the blue whale, the largest animal that has ever lived. After a peak kill of 29,606 in 1931, there was a steady decline to only 613 in the last season before the blue whale was given international protection in 1966. Once there may have been 200,000 blue whales; the maximum current worldwide estimate is barely 3,000. Some biologists doubt that the species can survive. The threatened extermination of the blue whale might be viewed as the culmination of a process that began in the Pleistocene: Human hunters, acting solely on the basis of immediate need, have systematically eliminated the large mammals of the earth.

Fortunately, during the 1930s a number of critically threatened species of whales were jointly protected by several nations. The most notable success of these measures was the recovery of gray whales along the west coast of North America. In 1946, the International Whaling Commission, a regulatory body now made up of 40 member nations, was formed. Although the commission has totally protected some species and set quotas on others, its effectiveness has been reduced by lack of cooperation from nations that have continued excessive harvests.

Concern for whales and other sea mammals led to enactment of the United States Marine Mammal Protection Act of 1972. This law basically prohibits the killing of marine mammals. But it allows numerous exceptions.

One controversy involves the sea otter. This species was pursued mercilessly by fur hunters in the 18th and 19th centuries. Barely 2,000 remained in 1911 when international protection was established. Today about 120,000 live in Alaskan waters alone. There, and in California, sea otters are being accused of depleting fisheries. Some people consider a harvest of them practical.

Valuable furbearing mammals have played an important role in human history. The search for pelts drew Russian adventurers across Siberia and into Alaska during the 1700s. Earlier, French explorers had penetrated the heart of North America and started to establish fur-trade routes. With the same objective, the English chartered the Hudson's Bay Company in 1670. Much of the Anglo-French rivalry on the continent over the next century resulted from competition for the fur trade. After the United States gained independence, the lure of furs stimulated westward movement.

As beavers became scarce in the West by the 1860s, trappers began to concentrate on the central plains. In such open habitat, they used not traps but poison to take wolves and other carnivorous furbearers. Early in the 20th century some eastern lowlands were discovered to be rich in furs. Surprisingly, the leading fur-producing area of the entire continent became—and continues to be—the state of Louisiana. The industry there was built primarily around the muskrat populations in the vast coastal marshes, but by 1962 the nutria, a rodent introduced from South America, was yielding greater profits.

Historically, the fur trade has depended upon such variables as economic conditions, fashions, and the population levels of target mammals. By the 1900s, trapping and habitat disruption had eliminated the beaver, marten, and fisher over most of the eastern United States. The river otter had been exterminated in many areas. Small furbearers, such as the mink, seemed destined to disappear in some states. Indeed, the sea mink of New England already had been eradicated.

Eskimos with flensing tools slice a bowhead whale to ribbons at Barrow, Alaska. Though endangered, the species may be legally hunted by aboriginal Americans. They follow ancestral ways as they communally harvest the whale. But to kill it they lay aside their ivory-tipped harpoons and pick up their guns.

In the 1930s and 1940s, history began to favor the fur trade's targets. Conservationists had won regulations putting many furbearers and their habitats under some kind of protection. The Depression and World War II slackened demand for luxurious furs. Most furbearers managed to maintain themselves or even make moderate recoveries. The most spectacular comeback has been the beaver's. It has reoccupied or has been reintroduced into much of its former range.

Some conservationists are still concerned about the impact of fur trapping, especially on mammals not covered by detailed regulations. In recent years there has been a general rise in prices for pelts, and a correspondingly greater kill of some species.

An increased harvest of bobcat fur came on top of large kills for sport and predator control—even though there were authoritative reports of seriously declining populations. Since most bobcat pelts were sold abroad, the United States government began to impose restrictions under an international trade convention that currently involves 128 nations. This agreement attempts to control commerce in many kinds of endangered wildlife, including all species of cats. Some states have improved research and management programs dedicated to the bobcat. But the controversy remains: How much protection does the bobcat need?

Furbearers are not the only animals preyed upon by the ultimate mammal. Hunters for the skin trade killed untold numbers of deer and bison. Indeed, the production of buffalo robes was a major industry during the 19th century. Large mammals were also hunted in great numbers for food and for sport as the frontier advanced. In the 16th century, the bison may have numbered 60 million, inhabiting not only the Great Plains but also the East as far as New York and Florida. By the 1830s the bison had been exterminated east of the Mississippi. By 1890 barely 1,000 survived, mostly in Canada. Subsequent conservation efforts sometimes are hailed as having saved the species. But in terms of the

continent-wide ecological role it once played, the bison may be gone forever.

Declines of other big game mammals in North America were nearly as catastrophic. The pronghorn, once almost as abundant as the bison, suffered a comparable fate as the plains were settled in the 19th century. Even before the American Revolution, the white-tailed deer had become so rare in the East that some colonies enacted protective laws. By 1900 there was fear for the survival of both

example, the number of pronghorns in the United States was estimated at 26,000; today the estimate is nearly one million. In the same period, white-tailed deer are believed to have increased from 500,000 to 14 million. What were described as mere "scattered herds" of mule deer now is an estimated population of five million. The elk's comeback has been estimated to be from 41,000 in 1900 to nearly one million today. Moose, nearly wiped out in 1900, now are up to about 25,000 in the lower

Locked moose antlers, remnant of a deadly duel, provide a red fox with a calcium-rich snack. Small mammals often feed on discarded antlers, for nothing is wasted in the food web of a natural environment. Here at Isle Royale National Park, natural controls still govern. Wolves cull moose, helping to keep the herd at a level the island can support.

that species and the related mule deer of the West. By the same time elk had disappeared in the eastern, central, and southwestern United States, and moose survived only in a few remote northern areas.

State laws and management programs, supported by federal funds and habitat protection, have reversed the disastrous trends of the previous century. Most of the big game species have been restored to substantial portions of their historic ranges. In 1900, for

48 states. There are at least another 145,000 moose in Alaska and 600,000 in Canada.

In much of the West, though, mule deer again are declining, partly because of habitat manipulation by people. In some areas, elk are threatened by oil exploration, agriculture, and logging. Yet, around the Great Lakes, white-tailed deer appear to have suffered because there is no longer enough logging to create favorable open habitat. Bighorn sheep and caribou seem more vulnerable to human interference than do most other big game species and do not respond as well to conservation efforts. Although some populations reportedly are thriving, there are fewer bighorns in the United States now than there were in the 1920s. For most of the last hundred years there has been a general downward trend in caribou numbers in Alaska and Canada. And the future of that species is clouded by the increasing tempo of oil and gas exploitation in the Arctic.

Killing for subsistence, sport, or profit—even when excessive—seldom involves dislike of the target mammals. But strong hatred flares in the warfare against carnivorous mammals that threaten livestock.

Colonial governments authorized bounty payments to persons killing wolves, bears, cougars, and other predators. The pattern repeated itself as the frontier advanced. The bounty system, however, not only was wasteful and destructive but also sometimes failed entirely. Small predators often bred fast enough to withstand the pressure. Large species usually collapsed only when the wilderness was broken up and their ranges became accessible to people.

Human extermination of deer and other natural prey species probably contributed to the decline of carnivores and may have stimulated their depredation on domestic animals. By the early 20th century, the cougar and gray wolf had been exterminated in the eastern United States, except for a few remote forests. The aggressive grizzly bear, always restricted to the West, was then rare in most stock-raising areas. The black bear, less obnoxious to people and more able to adapt to their presence, survived over much of the East, as well as in most of the West.

The United States government began its own predator-control campaign in 1915. The killing was fostered by pressure from the livestock industry and concern about the predatory mammals that still inhabited the newly formed national forests of the West. The gray wolf was the first major target, and by 1930 that species was practically a thing of the past in the West. Government wolf-control measures later were extended to Alaska, and emulated by Canada, mainly to protect caribou and introduced reindeer herds.

The most consistently and heavily hunted predator of this century has been the coyote. The federal kill often exceeded 100,000 per year. Far more have been killed by farmers, state and local agents, bounty hunters, sportsmen, and trappers. The spreading of poison bait was long a favorite means of killing coyotes and other predators. Conservationists fighting that method argued that many nontarget species were being decimated by indiscriminate poisoning. Finally, a 1972 presidential order banned the general use of poison for predator control.

Are too many cougars being killed to protect livestock in the Southwest? Should wolf numbers be controlled to increase numbers of caribou in Alaska and deer in southeastern Canada? Can the grizzly populations of the Yellowstone region and western Montana—the last in the lower 48 states—be maintained in the face of increasing commercial development and recreational activity? Perhaps all such questions boil down to this: Is there room for more than one large meat-eating mammal on the continent?

Large carnivores have not been the only mammals labeled as public enemies. Official poisoning operations have been directed against meadow voles that damaged fruit trees, muskrats that burrowed into dikes, porcupines that destroyed timber—and a host of other rodents. The most widespread campaigns have been aimed at eradicating ground squirrels and prairie dogs because they allegedly fed on crops and pasture.

The reduction of prairie dog numbers on the Great Plains nearly brought about the extinction of a species dependent upon them for prey—the black-footed ferret. Such indirect impacts, now increasing through growth of human population and technology, constitute the most serious problems facing American wildlife. Environmental disruption may hurt nearly all mammals, but the most

Research biologist David Mech approaches his subject with a syringe on a stick. A wolf is first caught in a foot trap. Then, after an injection of a tranquilizer takes effect, the wolf is fitted with a collar containing a radio for tracking the animal's movements. Mech (an author in this book) then carries the wolf to a secluded spot for recovery. Such studies lead to better protection for this vanishing species.

OVERLEAF: *Regally silhouetted in Yellowstone National Park, a lordly moose wades gilded shallows. Once nearly gone from the United States, the moose now gets protection from the ultimate mammal.*

susceptible are small species, which have strict limits on their habitats or food supplies.

Great Gull Island, off Long Island, New York, once had its own species of vole. During the Spanish-American War, amid fears that the East Coast would be attacked, fortifications were built on the island. The vole's tiny habitat was obliterated, and the species has not been seen since.

During the winter of 1972-1973, specimens that resembled the widespread marsh rice rat *Oryzomys palustris*—but with a unique silvery pelage—were found on Cudjoe Key at the southern tip of Florida. In the November 1978 *Journal of Mammalogy* the rodent was formally named *Oryzomys argentatus*—the silver rice rat—and it thus became a newly recognized species of mammal in North America. Later the species was discovered on other small islands, but at a much lower population density than *O. palustris*. Commercial development is taking the limited habitat of the silver rice rat; competition and predation from introduced rats jeopardize it. And so one of our newer mammals is also among our most endangered. Many other species of rodents, insectivores, and bats are jeopardized in Florida and other rapidly developing parts of the Sunbelt.

Should we worry that these small creatures may pass from the scene? Is not extinction a normal process that will overtake them anyway? When people occupy an area and use it to their advantage are they not acting the same as any other mammal?

Extinction is a natural process. So is the death of an individual. But this does not mean murder should be condoned. People have the ability not only to modify the environment but also to find and implement means of conservation. While we may never again have room for great herds of wild bison, we can preserve all the habitat required by dozens of smaller species. We therefore can insure that at least some of our native mammals, living as they should, under fully natural ecological and behavioral conditions, will remain with us.   RONALD M. NOWAK